Instructor's Manual
with Software and Transparency Masters

Engineering
Problem Solving
with ANSI C:
Fundamental Concepts

Delores M. Etter

Prentice-Hall, Inc., Englewood Cliffs, NJ 07632

© 1995 by Prentice-Hall, Inc.
A Simon & Schuster Company
Englewood Cliffs, N.J. 07632

10 9 8 7 6 5 4 3 2

ISBN 0-13-310814-7
Printed in the United States of America

Contents

Preface

This Instructor's Manual contains solutions to all the end-of-chapter problems in *Engineering Problem Solving with ANSI C: Fundamental Concepts*. The diskette that accompanies this Instructor's Manual contains the ASCII files for these end-of-chapter problem solutions. The `readme` file describes the organization of the files on the diskette. Any data files referenced in the programs are included on the diskette that accompanies the text.

A set of 50 transparency masters are also included in this Instructor's Manual. The contents of these transparency masters, chosen from figures and diagrams from the text, are listed below:

1. Internal organization of a computer
2. Software interface to the computer
3. Comparison of software statements
4. Program compilation/linking/execution
5. Software life-cycle phases
6. Variable declaration and initialization
7. Keywords
8. Numeric data types
9. Example data-type limits
10. Precedence of arithmetic operators
11. Precedence of arithmetic and assignment operators
12. Numeric conversion specifiers for output statements
13. Examples with conversion specifiers
14. Escape sequences
15. Numeric conversion specifiers for input statements
16. Linear and cubic spline interpolation
17. Linear interpolation using similar triangles
18. Pseudocode notation and flowchart symbols
19. Example of a flowchart for a sequence structure
20. Example of a flowchart for a selection structure
21. Example of a flowchart for a repetition structure
22. Relational operators
23. Logical operators
24. Operator precedence for arithmetic, relational, and logical operators
25. Distances from a linear estimate to a set of points
26. Equations for the "best" linear fit in terms of least squares

All of the programs have been computer tested, but some errors always manage to slip through, both in the text and in the Instructor's Manual. I would appreciate it very much if you would share with me any errors that you find so that they can be corrected when the text or this Instructor's Manual are reprinted.

Professor Delores M. Etter
Electrical/Computer Engr. Dept.
Campus Box 425
University of Colorado
Boulder, CO   80309-0425
etter@boulder.colorado.edu

Chapter 2

```
/*-----------------------------------------------------------------*/
/*   Problem chapter2_1                                            */
/*                                                                 */
/*   This program converts miles to kilometers.                   */

#include <stdio.h>
#include <stdlib.h>

main()
{
    /*  Declare variables.  */
    double miles, kilometers;

    /*  Enter number of miles from the keyboard.  */
    printf("Enter the number of miles: \n");
    scanf("%lf",&miles);

    /*  Compute the number of kilometers equal to the specified miles. */
    kilometers = 1.6093440*miles;

    /*  Print the number of kilometers.  */
    printf("%8.3f miles = %8.3f kilometers \n",miles,kilometers);

    /*  Exit program.  */
    return EXIT_SUCCESS;
}
/*-----------------------------------------------------------------*/

/*-----------------------------------------------------------------*/
/*   Problem chapter2_2                                            */
/*                                                                 */
/*   This program converts meters to miles.                       */

#include <stdio.h>
#include <stdlib.h>

main()
{
    /*  Declare variables.  */
    double miles, meters;

    /*  Enter number of meters from the keyboard.  */
    printf("Enter the number of meters: \n");
    scanf("%lf",&meters);

    /*  Compute the number of miles equal to the specified meters.  */
    miles = meters/1609.3440;

    /*  Print the number of miles.  */
    printf("%8.3f meters = %8.3f miles \n",meters,miles);

    /*  Exit program.  */
    return EXIT_SUCCESS;
}
/*-----------------------------------------------------------------*/

/*-----------------------------------------------------------------*/
```

```
/*   Problem chapter2_3                                                    */
/*                                                                         */
/*   This program converts pounds to kilograms.                           */

#include <stdio.h>
#include <stdlib.h>

main()
{
    /*   Declare variables.   */
    double pounds, kilograms;

    /*   Enter number of pounds from the keyboard.   */
    printf("Enter the number of pounds: \n");
    scanf("%lf",&pounds);

    /*   Compute number of kilograms equal to the specified pounds.   */
    kilograms = pounds/2.205;

    /*   Print the number of kilograms.   */
    printf("%8.3f pounds = %8.3f kilograms \n",pounds,kilograms);

    /*   Exit program.   */
    return EXIT_SUCCESS;
}
/*-------------------------------------------------------------------*/

/*-------------------------------------------------------------------*/
/*   Problem chapter2_4                                                    */
/*                                                                         */
/*   This program converts newtons to pounds.                             */

#include <stdio.h>
#include <stdlib.h>

main()
{
    /*   Declare variables.   */
    double pounds, newtons;

    /*   Enter number of newtons from the keyboard.   */
    printf("Enter the number of newtons:  \n");
    scanf("%lf",&newtons);

    /*   Compute number of pounds equal to the specified newtons.   */
    pounds = newtons/4.448;

    /*   Print the number of pounds.   */
    printf("%8.3f newtons = %8.3f pounds \n",newtons,pounds);

    /*   Exit program.   */
    return EXIT_SUCCESS;
}
/*-------------------------------------------------------------------*/

/*-------------------------------------------------------------------*/
/*   Problem chapter2_5                                                    */
/*                                                                         */
/*   This program converts degrees Fahrenheit to degrees Rankin.          */
```

```c
#include <stdio.h>
#include <stdlib.h>

main()
{
    /*  Declare variables.  */
    double degrees_F, degrees_R;

    /*  Enter temperture in degrees Fahrenheit from the keyboard.  */
    printf("Enter the temperature in degrees Fahrenheit:  \n");
    scanf("%lf",&degrees_F);

    /*  Compute the equivalent temperature in degrees Rankin  */
    /*   from the given temperature.                          */
    degrees_R = degrees_F + 459.67;

    /*  Print the temperatures.  */
    printf("%8.3f degrees Fahrenheit = %8.3f degrees Rankin \n",
           degrees_F,degrees_R);

    /*  Exit program.  */
    return EXIT_SUCCESS;
}
/*------------------------------------------------------------------*/


/*------------------------------------------------------------------*/
/*   Problem chapter2_6                                             */
/*                                                                  */
/*   This program converts degrees Celsius to degrees Rankin.       */

#include <stdio.h>
#include <stdlib.h>

main()
{
    /*  Declare variables.  */
    double degrees_C, degrees_R, degrees_F;

    /*  Enter temperture in degrees Celsius from the keyboard.  */
    printf("Enter the temperature in degrees Celsius:  \n");
    scanf("%lf",&degrees_C);

    /*  Compute the equivalent temperature in degrees Rankin  */
    /*   from the given temperature.                          */
    degrees_F = (9.0/5.0)*degrees_C + 32;
    degrees_R = degrees_F + 459.67;

    /*  Print the temperatures.  */
    printf("%8.3f degrees Celsius = %8.3f degrees Rankin \n",
           degrees_C,degrees_R);

    /*  Exit program.  */
    return EXIT_SUCCESS;
}
/*------------------------------------------------------------------*/


/*------------------------------------------------------------------*/
/*   Problem chapter2_7                                             */
/*                                                                  */
/*   This program converts degrees Kelvin to degrees Fahrenheit.    */
```

3

```c
#include <stdio.h>
#include <stdlib.h>

main()
{
    /*  Declare variables.  */
    double degrees_R, degrees_K, degrees_F;

    /*  Enter temperture in degrees Kelvin from the keyboard.  */
    printf("Enter the temperature in degrees Kelvin:  \n");
    scanf("%lf",&degrees_K);

    /*  Compute the equivalent temperature in degrees Fahrenheit  */
    /*   from the given temperature.                              */
    degrees_R = (9.0/5.0)*degrees_K;
    degrees_F = degrees_R - 459.67;

    /*  Print the temperatures.  */
    printf("%8.3f degrees Kelvin = %8.3f degrees Fahrenheit \n",
            degrees_K,degrees_F);

    /*  Exit program.  */
    return EXIT_SUCCESS;
}
/*-------------------------------------------------------------------*/


/*-------------------------------------------------------------------*/
/*  Problem chapter2_8                                             */
/*                                                                */
/*  This program finds the area of a rectangle.                   */

#include <stdio.h>
#include <stdlib.h>

main()
{
    /*  Declare variables.  */
    double a, b, area;

    /*  Enter the lengths of sides of the rectangle.  */
    printf("Enter the lengths of the sides of the rectangle:  \n");
    scanf("%lf %lf",&a,&b);

    /*  Compute the area of the rectangle.  */
    area = a*b;

    /*  Print the value of the area.  */
    printf("The area of a rectangle with sides %5.3f and %5.3f"
            " is %5.3f. \n",a,b,area);

    /*  Exit program.  */
    return EXIT_SUCCESS;
}
/*-------------------------------------------------------------------*/


/*-------------------------------------------------------------------*/
/*  Problem chapter2_9                                            */
/*                                                               */
/*  This program finds the area of a triangle.                   */
```

4

```c
#include <stdio.h>
#include <stdlib.h>

main()
{
    /*  Declare variables.  */
    double h, b, area;

    /*  Enter the base and the height of the triangle.  */
    printf("Enter the base and the height of the triangle: \n");
    scanf("%lf %lf",&b,&h);

    /*  Compute the area of the triangle.  */
    area = 0.5*b*h;

    /*  Print the value of the area.  */
    printf("The area of a triangle with base %5.3f and height %5.3f "
           "is %5.3f. \n",b,h,area);

    /*  Exit program.  */
    return EXIT_SUCCESS;
}
/*-------------------------------------------------------------------*/

/*-------------------------------------------------------------------*/
/*  Problem chapter2_10                                              */
/*                                                                   */
/*  This program finds the area of a circle.                         */

#include <stdio.h>
#include <stdlib.h>
#define PI 3.141593

main()
{
    /*  Declare variables.  */
    double r, area;

    /*  Enter the radius.  */
    printf("Enter the radius of the circle: \n");
    scanf("%lf",&r);

    /*  Compute the area of the circle.  */
    area = PI*r*r;

    /*  Print the value of the area.  */
    printf("The area of a cirlce with radius %5.3f is  %5.3f. \n",
           r,area);

    /*  Exit program.  */
    return EXIT_SUCCESS;
}
/*-------------------------------------------------------------------*/

/*-------------------------------------------------------------------*/
/*  Problem chapter2_11                                              */
/*                                                                   */
/*  This program computes the area of a sector of a circle when      */
/*  theta is the angle in radians between the radii.                 */
```

```c
#include <stdio.h>
#include <stdlib.h>

main()
{
    /*  Declare variables.  */
    double theta, r, area;

    /*  Enter the lengths of the radii and */
    /*   the angle between them.           */
    printf("Enter the length of the radii and the angle "
           "(in radians) between them: \n");
    scanf("%lf %lf",&r,&theta);

    /*  Compute the area of the sector.  */
    area = (r*r*theta)/2.0;

    /*  Print the value of the area.  */
    printf("The area of sector is %5.3f. \n",area);

    /*  Exit program.  */
    return EXIT_SUCCESS;
}
/*------------------------------------------------------------------*/


/*------------------------------------------------------------------*/
/*  Problem chapter2_12                                             */
/*                                                                  */
/*  This program computes the area of a sector of a circle when     */
/*   d is the angle in degrees between the radii.                   */

#include <stdio.h>
#include <stdlib.h>
#define PI 3.141593

main()
{
    /*  Declare variables.  */
    double d, theta, r, area;

    /*  Enter the length of the radii and the angle between them.  */
    printf("Enter the length of the radii and the angle"
           " (in degrees) between them: \n");
    scanf("%lf %lf",&r,&d);

    /*  Compute the area of the sector.  */
    theta = (d*PI)/180;
    area = (r*r*theta)/2.0;

    /*  Print the value of the area.  */
    printf("The area of sector is %5.3f. \n",area);

    /*  Exit program.  */
    return EXIT_SUCCESS;
}
/*------------------------------------------------------------------*/


/*------------------------------------------------------------------*/
/*  Problem chapter2_13                                             */
```

```
/*
/*  This program computes the area of an
/*  ellipse with semiaxes a and b.

#include <stdio.h>
#include <stdlib.h>

#define PI 3.141593

main()
{
    /*  Declare variables.  */
    double a, b, area;

    /*  Enter the length of the semiaxes.  */
    printf("Enter the length of the semiaxes: \n");
    scanf("%lf %lf",&a,&b);

    /*  Compute the area of the ellipse.  */
    area = PI*a*b;

    /*  Print the value of the area.  */
    printf("The area of an ellipse with semiaxes %5.3f and %5.3f"
           " is %5.3f. \n",a,b,area);

    /*  Exit program.  */
    return EXIT_SUCCESS;
}
/*-------------------------------------------------------------------*/

/*-------------------------------------------------------------------*/
/*  Problem chapter2_14                                              */
/*                                                                   */
/*  This program computes the area of the surface                    */
/*  of a sphere of radius r.                                         */

#include <stdio.h>
#include <stdlib.h>
#define PI 3.141593

main()
 {
   /*  Declare variables.  */
   double r, area;

   /*  Enter the radius of the sphere.  */
   printf("Enter the radius of the sphere: \n");
   scanf("%lf",&r);

   /*  Compute the area of the sphere.  */
   area = 4.0*PI*r*r;

   /*  Print the value of the area.  */
   printf("The area of a sphere with radius %5.3f"
          " is %5.3f. \n",r,area);

   /*  Exit program.  */
   return EXIT_SUCCESS;
}
/*-------------------------------------------------------------------*/
```

```
/*-----------------------------------------------------------------*/
/*   Problem chapter2_15                                           */
/*                                                                 */
/*   This program computes the volume of a sphere of radius r.     */

#include <stdio.h>
#include <stdlib.h>
#define PI 3.141593

main()
{
    /*  Declare variables.  */
    double r, volume;

    /*  Enter the radius of the sphere.  */
    printf("Enter the radius of the sphere: \n");
    scanf("%lf",&r);

    /*  Compute the volume of the sphere.  */
    volume = (4.0/3.0)*PI*r*r*r;

    /*  Print the volume.  */
    printf("The volume of a sphere with radius %5.3f is %5.3f. \n",
           r,volume);

    /*  Exit program.  */
    return EXIT_SUCCESS;
}
/*-----------------------------------------------------------------*/

/*-----------------------------------------------------------------*/
/*   Problem chapter2_16                                           */
/*                                                                 */
/*   This program computes the volume of a cylinder               */
/*   of radius r and height h.                                     */

#include <stdio.h>
#include <stdlib.h>
#define PI 3.141593

main()
{
    /*  Declare variables.  */
    double r, h, volume;

    /*  Enter the radius and height of the cylinder.  */
    printf("Enter the radius and the height of the cylinder: \n");
    scanf("%lf %lf",&r,&h);

    /*  Compute the volume of the cylinder.  */
    volume = PI*r*r*h;

    /*  Print the volume.  */
    printf("The volume of a cylinder of radius %5.3f and "
           "height %5.3f is %5.3f. \n",r,h,volume);

    /*  Exit program.  */
    return EXIT_SUCCESS;
}
```

```
/*-------------------------------------------------------------------*/

/*-------------------------------------------------------------------*/
/*   Problem chapter2_17                                           */
/*                                                                 */
/*   This program computes the molecular weight of the            */
/*   amino acid glycine.                                          */

#include <stdio.h>
#include <stdlib.h>

/*   Defines symbolic constants for the appropriate atomic weights.   */
#define OXYGEN 15.9994
#define CARBON 12.011
#define NITROGEN 14.00674
#define HYDROGEN 1.00794

main()
{
    /*   Declare variables.   */
    double molecular_weight;

    /*   Compute the molecular weight of glycine.   */
    molecular_weight = (2*OXYGEN) + (2*CARBON) +
                       NITROGEN + (5*HYDROGEN);

    /*   Print the molecular weight.   */
    printf("The molecular weight of glycine is %4.3f. \n",
           molecular_weight);

    /*   Exit program.   */
    return EXIT_SUCCESS;
}
/*-------------------------------------------------------------------*/

/*-------------------------------------------------------------------*/
/*   Problem chapter2_18                                           */
/*                                                                 */
/*   This program computes the molecular weights of the           */
/*   amino acid glutamic and glutamine.                           */

#include <stdio.h>
#include <stdlib.h>

/*   Defines symbolic constants for the appropriate atomic weights.   */
#define OXYGEN 15.9994
#define CARBON 12.011
#define NITROGEN 14.00674
#define HYDROGEN 1.00794

main()
{
    /*   Declare variables.   */
    double molecular_weight1, molecular_weight2;

    /*   Compute the molecular weights of glutamic and glutamine.   */
    molecular_weight1 = (4*OXYGEN) + (5*CARBON) +
                        NITROGEN + (8*HYDROGEN);
    molecular_weight2 = (3*OXYGEN) + (5*CARBON) +
                        (2*NITROGEN) + (10*HYDROGEN);
```

```c
     /*  Print the molecular weights.  */
     printf("The molecular weight of glutamic is %4.3f. \n",
             molecular_weight1);
     printf("The molecular weight of glutamine is %4.3f. \n",
             molecular_weight2);

     /*  Exit program.  */
     return EXIT_SUCCESS;
}
/*-----------------------------------------------------------------*/

/*-----------------------------------------------------------------*/
/*  Problem chapter2_19                                            */
/*                                                                 */
/*  This program computes the molecular weight of a particular     */
/*  amino acid given the number of atoms for each of the five      */
/*  elements found in the amino acid.                              */

#include <stdio.h>
#include <stdlib.h>

/*  Defines symbolic constants for the appropriate atomic weights.  */
#define OXYGEN 15.9994
#define CARBON 12.011
#define NITROGEN 14.00674
#define HYDROGEN 1.00794
#define SULFUR 32.066

main()
{
     /*  Declare variable.  */
     int no_oxy, no_carbon, no_nitro, no_hydro, no_sulfur;
     double molecular_weight;

     /*  Enter the number of atoms for each of the five elements.  */
     printf("Enter the number of oxygen atoms found "
             "in the amino acid. \n");
     scanf("%i",&no_oxy);
     printf("Enter the number of carbon atoms. \n");
     scanf("%i",&no_carbon);
     printf("Enter the number of nitrogen atoms. \n");
     scanf("%i",&no_nitro);
     printf("Enter the number of sulfur atoms. \n");
     scanf("%i",&no_sulfur);
     printf("Enter the number of hydrogen atoms. \n");
     scanf("%i",&no_hydro);

     /*  Compute the molecular weight.  */
     molecular_weight = (no_oxy*OXYGEN) + (no_carbon*CARBON) +
                         (no_nitro*NITROGEN) + (no_sulfur*SULFUR) +
                         (no_hydro*HYDROGEN);

     /*  Print the molecular weight.  */
     printf("The molecular weight of this particular amino acid"
             " is %4.3f. \n",molecular_weight);

     /*  Exit program.  */
     return EXIT_SUCCESS;
}
```

```
/*-------------------------------------------------------------------*/

/*-------------------------------------------------------------------*/
/*  Problem chapter2_20                                            */
/*                                                                 */
/*  This program computes the average atomic weight of the atoms   */
/*  found in a particular amino acid given the number of atoms for */
/*  each of the five elements found in amino acid.                 */

#include <stdio.h>
#include <stdlib.h>

/*  Defines symbolic constants for the appropriate atomic weights.  */
#define OXYGEN 15.9994
#define CARBON 12.011
#define NITROGEN 14.00674
#define HYDROGEN 1.00794
#define SULFUR 32.066

main()
{
    /*  Declare variables.  */
    int no_oxy, no_carbon, no_nitro, no_hydro, no_sulfur, total_no;
    double average_atomic_weight;

    /*  Enter the number of atoms for each of the five elements.  */
    printf("Enter the number of oxygen atoms found "
            " in the amino acid. \n");
    scanf("%i",&no_oxy);
    printf("Enter the number of carbon atoms. \n");
    scanf("%i",&no_carbon);
    printf("Enter the number of nitrogen atoms. \n");
    scanf("%i",&no_nitro);
    printf("Enter the number of sulfur atoms. \n");
    scanf("%i",&no_sulfur);
    printf("Enter the number of hydrogen atoms. \n");
    scanf("%i",&no_hydro);

    /*  Compute the average weight of the atoms.  */
    total_no = no_oxy + no_carbon + no_nitro + no_sulfur + no_hydro;
    average_atomic_weight = ((no_oxy*OXYGEN) + (no_carbon*CARBON) +
                             (no_nitro*NITROGEN) + (no_sulfur*SULFUR) +
                             (no_hydro*HYDROGEN))/total_no;

    /*  Print the average atomic weight.  */
    printf("The average weight of the atoms in this particular amino "
            "acid is %4.3f. \n",average_atomic_weight);

    /*  Exit program.  */
    return EXIT_SUCCESS;
}
/*-------------------------------------------------------------------*/

/*-------------------------------------------------------------------*/
/*  Problem chapter2_21                                            */
/*                                                                 */
/*  This program reads in a positive number and then computes      */
/*  the logarithm of that value to the base 2.                     */

#include <stdio.h>
```

```c
#include <stdlib.h>
#include <math.h>

main()
{
    /*  Declare variables.  */
    double  x, answer;

    /*  Enter a positive number.  */
    printf("Enter a positive number:  \n");
    scanf("%lf",&x);

    /*  Compute the logarithm to base 2.  */
    answer = log(x)/log(2.0);

    /*  Print the answer.  */
    printf("The logarithm of %5.3f to the base 2 is "
           " %5.3f. \n",x,answer);

    /*  Exit program.  */
    return EXIT_SUCCESS;
}
/*------------------------------------------------------------------*/


/*------------------------------------------------------------------*/
/*   Problem chapter2_22                                            */
/*                                                                  */
/*   This program reads in a positive number and then computes      */
/*   the logarithm of that value to the base 8.                     */

#include <stdio.h>
#include <stdlib.h>
#include <math.h>

main()
{
    /*  Declare variables.  */
    double  x, answer;

    /*  Enter a positive number.  */
    printf("Enter a positive number:  \n");
    scanf("%lf",&x);

    /*  Compute the logarithm to base 8.  */
    answer = log(x)/log(8.0);

    /*  Print the answer.  */
    printf("The logarithm of %5.3f to the base 8 is %5.3f. \n",
           x,answer);

    /*  Exit program.  */
    return EXIT_SUCCESS;
}
/*------------------------------------------------------------------*/
```

Chapter 3

```
/*------------------------------------------------------------------*/
/*   Problem chapter3_1                                             */
/*                                                                  */
/*   This program prints a conversion table from radians to degrees. */
/*   The radian column starts at 0.0, and increments by pi/10,      */
/*   until the radian amount is 2*pi.                               */

#include <stdio.h>
#include <stdlib.h>
#define PI 3.141593

main()
{
    /*  Declare and initialize variables.  */
    double degrees, increment=PI/10, radians=10;

    /*  Print radians and degrees in a loop. */
    printf("Radians to Degrees \n");
    while (radians <= 2*PI)
    {
        degrees = radians*180/PI;
        printf("%9.6f  %9.6f \n",radians,degrees);
        radians = radians + increment;
    }

    /*  Exit program.  */
    return EXIT_SUCCESS;
}
/*------------------------------------------------------------------*/

/*------------------------------------------------------------------*/
/*   Problem chapter3_2                                             */
/*                                                                  */
/*   This program prints a conversion table from degrees to radians. */
/*   The first line contains the value for zero degrees and the     */
/*   last line contains the value for 360 degrees. The user enters  */
/*   the increment to use between the lines in the table.           */

#include <stdio.h>
#include <stdlib.h>
#define PI 3.141593

main()
{
    /*  Declare and initialize variables.  */
    double degrees=0, increment=0, radians=0;

    /*  Prompt user for increment value.  */
    printf("Enter value for increment between lines:");
    scanf("%lf",&increment);

    /*  Print title and table  */
    printf("Degrees to Radians with increment %9.6f \n",increment);
    while (degrees < 360 )
    {
        radians = degrees *PI/180;
        printf("%9.6f  %9.6f\n",degrees,radians);
```

```
            degrees += increment;
        }

        /*  Print a value for 360 degrees.  */
        degrees = 360;
        radians = degrees*PI/180;
        printf("%9.6f  %9.6f\n",degrees,radians);

        /*  Exit program.  */
        return EXIT_SUCCESS;
}
/*-------------------------------------------------------------------*/

/*-------------------------------------------------------------------*/
/*   Problem chapter3_3                                              */
/*                                                                   */
/*   This program prints a conversion table for inches to centimeters. */
/*   The inches column starts at 0.0 and increments by 0.5 in.       */
/*   The last line contains the value 20.0 in.                       */

#include <stdio.h>
#include <stdlib.h>
#define CM_PER_INCH 2.54

main()
{
        /*  Declare and initialize variables.  */
        double cm=0.0, inches=0.0, increment=0.5;

        /*  Print table title and values.  */
        printf("Inches to Centimeters\n");
        while (inches <= 20.0)
        {
            cm = inches*CM_PER_INCH;
            printf("%9.6f  %9.6f\n",inches,cm);
            inches += increment;
        }

        /*  Exit program.  */
        return EXIT_SUCCESS;
}
/*-------------------------------------------------------------------*/

/*-------------------------------------------------------------------*/
/*   Problem chapter3_4                                              */
/*                                                                   */
/*   This program generates a conversion table from mph to ft/s.     */
/*   It starts the mph column at 0, and increment by 5 mph.          */
/*   The last line contains the value 65 mph.                        */

#include <stdio.h>
#include <stdlib.h>
#define FT_PER_MI 5280
#define SEC_PER_HOUR 3600

main()
{
        /* Define and initialize variables.  */
        int increment=5, mph=0;
        double fps=0;
```

```c
    /*  Print titles and table.  */
    printf("Miles/hour to Feet/second\n");
    while (mph <= 65)
    {
        fps = (double)mph*FT_PER_MI/SEC_PER_HOUR;
        printf("%i  %f\n",mph,fps);
        mph += increment;
    }

    /*  Exit successfully.  */
    return EXIT_SUCCESS;
}
/*-------------------------------------------------------------------*/


/*-------------------------------------------------------------------*/
/*  Problem chapter3_5                                               */
/*  This program generates a conversion table from ft/s to mph.      */
/*  It starts the ft/s column at 0, and increments by 5 ft/s.        */
/*  The last line contains the value 100 ft/s.                       */

#include <stdio.h>
#include <stdlib.h>
#define FT_PER_MI 5280
#define SEC_PER_HOUR 3600

main()
{
    /*  Define and initialize variables.  */
    int fps=0, increment=5;
    double mph=0;

    /*  Print titles and table.  */
    printf("Feet/second to Miles/hour \n");
    while (fps <= 100)
    {
        mph = (double)fps*SEC_PER_HOUR/FT_PER_MI;
        printf("%i  %f\n",fps,mph);
        fps += increment;
    }

    /*  Exit program.  *.
    return EXIT_SUCCESS;
}
/*-------------------------------------------------------------------*/


/*-------------------------------------------------------------------*/
/*  Problem chapter3_6                                               */
/*                                                                  */
/*  This program generates a table of conversions from francs to    */
/*  dollars. The table starts with the francs column at 5 Fr.       */

#include <stdio.h>
#include <stdlib.h>
#define FR_PER_DOLLAR 5.3

main()
{
    /*  Define and initialize variables.  */
    int francs=5, increment=5;
```

```
        double dollars;

        /*  Print title and table.  */
        printf("French Francs to US Dollars \n");
        for (francs=5; francs<=125; francs += increment)
        {
            dollars = (double)francs/FR_PER_DOLLAR;
            printf("%i Fr   $%6.2f \n",francs,dollars);
        }

        /*  Exit program.  */
        return EXIT_SUCCESS;
}
/*--------------------------------------------------------------------*/

/*--------------------------------------------------------------------*/
/*  Problem chapter3_7                                                */
/*                                                                    */
/*  This program generates a table of conversions from deutsche      */
/*  marks to francs.  The deutsche marks column starts at 1 DM and   */
/*  increments by 2 DM.                                              */

#include <stdio.h>
#include <stdlib.h>
#define FR_PER_DOLLAR 5.3
#define DM_PER_DOLLAR 1.57

main()
{
        /*  Define and initialize variables.  */
        int dmarks=1, increment=2, loopcount;
        double francs;

        /*  Print title and table.  */
        printf("Deutsche Marks to French Francs\n");
        for (loopcount=1; loopcount<=30; loopcount++)
        {
            francs = (double)dmarks/DM_PER_DOLLAR*FR_PER_DOLLAR;
            printf("%i DM   %6.2f Fr \n",dmarks,francs);
            dmarks += increment;
        }

        /*  Exit program.  */
        return EXIT_SUCCESS;
}
/*--------------------------------------------------------------------*/

/*--------------------------------------------------------------------*/
/*  Problem chapter3_8                                                */
/*                                                                    */
/*  This program prints a conversion table for yen to deutsche marks. */
/*  The yen column starts at 100Y.  25 lines are printed with the    */
/*  final line containing 10,000Y.                                   */

#include <stdio.h>
#include <stdlib.h>
#define DOLLAR_PER_YEN 0.0079
#define DM_PER_DOLLAR 1.57

main()
```

16

```
{
    /*  Define and initialize variables.  */
    int loopcount;
    double increment=412.5;     /* (10,000 - 100)/24; */
    double dmarks, yen=100.0;

    /*  Print title and table.  */
    printf("Japanese Yen to Deustche Marks\n");
    for (loopcount=1; loopcount<=25; loopcount++)
    {
        dmarks = yen*DOLLAR_PER_YEN*DM_PER_DOLLAR;
        printf("%8.2f Y   %8.4f DM \n",yen,dmarks);
        yen += increment;
    }

    /*  Exit program.  */
    return EXIT_SUCCESS;
}
/*-------------------------------------------------------------------*/

/*-------------------------------------------------------------------*/
/*  Problem chapter3_9                                               */
/*                                                                   */
/*  This program prints a table of conversions for dollars to francs, */
/*  deutsche marks, and yen.  The dollars column starts with $1 and  */
/*  increments by $1. 50 lines are printed in the table.             */

#include <stdio.h>
#include <stdlib.h>
#define FR_PER_DOLLAR 5.3
#define DM_PER_DOLLAR 1.57
#define DOLLAR_PER_YEN 0.0079

main()
{
    /*  Define and initialize variables.  */
    int dollars=1, number_lines=50;
    double dmarks, francs, yen;

    /*   Print title and table.  */
    printf("US Dollars to Francs,Deutsche Marks,Yen \n");
    for (dollars=1; dollars <=number_lines; dollars++)
    {
        francs = (double)dollars*FR_PER_DOLLAR;
        dmarks = (double)dollars*DM_PER_DOLLAR;
        yen = (double)dollars/DOLLAR_PER_YEN;
        printf("$%i  %8.4fFr   %8.4fDM   %8.4fY \n",
                dollars,francs,dmarks,yen);
    }

    /*  Exit program.  */
    return EXIT_SUCCESS;
}
/*-------------------------------------------------------------------*/

/*-------------------------------------------------------------------*/
/*  Problem chapter3_10                                              */
/*                                                                   */
/*  This program generates a table of conversions from Fahrenheit   */
/*  Celsius for values from 0 degrees F to 100 degrees F.           */
```

```
#include <stdio.h>
#include <stdlib.h>

main()
{
    /*  Define and initialize variables.  */
    int farenheit=0, increment=5;
    double celsius;

    /*  Print title and table.  */
    printf("Farenheit to Celsius in 5-degree increments \n");
    while (farenheit <= 100)
    {
        celsius = (5.0/9.0)*((double)farenheit - 32 );
        printf("%i F   %4.2f  C \n",farenheit,celsius);
        farenheit += increment;
    }

    /*  Exit program.  */
    return EXIT_SUCCESS;
}
/*-----------------------------------------------------------------*/

/*-----------------------------------------------------------------*/
/*   Problem chapter3_11                                           */
/*                                                                 */
/*   This program generates a table of conversions from Fahrenheit */
/*   to Kelvin for values from 0 degrees F to 200 degrees F. It    */
/*   allows the user to  enter the increment between lines.        */

#include <stdio.h>
#include <stdlib.h>

main()
{
    /*  Define and initialize variables.  */
    double farenheit=0, increment=0, kelvin;

    /*  Prompt user for increment.  */
    while (increment <= 0)
    {
        printf("Enter increment for table:");
        scanf("%lf",&increment);
    }

    /*   Print title and table.  */
    printf("Farenheit to Kelvin \n");
    do
    {
        kelvin = (5.0/9.0)*(farenheit + 459.67);
        printf("%4.2f F   %4.2f K \n",farenheit,kelvin);
        farenheit += increment;
    } while (farenheit <= 200.0);

    /*  Exit program.  */
    return EXIT_SUCCESS;
}
/*-----------------------------------------------------------------*/
```

```
/*-----------------------------------------------------------------------*/
/*  Problem chapter3_12                                                   */
/*                                                                        */
/*  This program generates a table of conversions from Celsius to        */
/*  Rankin.  It allows the user to enter the starting temperature        */
/*  and the increment between lines.                                      */

#include <stdio.h>
#include <stdlib.h>

main()
{
    /*  Define and initialize variables.  */
    int loopcount, loop_max = 25;
    double celcius, farenheit, increment=0, rankin;

    /*  Prompt user for starting temperature and increment.  */
    printf("Enter starting temperature in Celsius: ");
    scanf("%lf",&celsius);
    while (increment <= 0)
    {
        printf("Enter increments in degrees Celsius: ");
        scanf("%lf",&increment);
    }

    /*   Print title and table.  */
    printf("Celsius to Rankin \n");
    for (loopcount=1; loopcount<=loop_max; loopcount++)
    {
        farenheit = (9.0/5.0) * celsius + 32;
        rankin = farenheit + 459.67;
        printf("%4.2f C   %4.2f R \n",celsius,rankin);
        celsius += increment;
    }

    /*  Exit program.  */
    return EXIT_SUCCESS;
}
/*-----------------------------------------------------------------------*/


/*-----------------------------------------------------------------------*/
/*  Problem chapter3_13                                                   */
/*                                                                        */
/*  This program assumes that the file rocket1.dat contains an           */
/*  initial lline that contains the number of actual datalines           */
/*  that follows.  The program reads these data and determines           */
/*  the time at which the rocket begins falling back to earth.           */

#include <stdio.h>
#include <stdlib.h>
#define FILENAME "rocket1.dat"

main()
{
    /*  Declare variables.  */
    int number_of_items;
    double acceleration, altitude=0, previous_altitude=0.0,
           previous_time, time=0, velocity;
    FILE *rocket1;
```

```
    /*  Open input file.  */
    rocket1=fopen(FILENAME,"r");
    if (fscanf(rocket1,"%i",&number_of_items) < 1)
    {
        printf("No data in file rocket1\n");
        return EXIT_FAILURE;
    }

    /* Print rocket information.  */
    while ((number_of_items-->0) && (altitude>=previous_altitude))
    {
        previous_altitude = altitude;
        previous_time = time;
        fscanf(rocket1,"%lf %lf %lf %lf",
               &time,&altitude,&velocity,&acceleration);
    }

    /*  Print time at which rocket begins to fall.  */
    if (altitude < previous_altitude)
        printf("Time at which the rocket begins to fall is between "
               "%4.2f seconds and %4.2f seconds \n",previous_time,time);
    else
        printf("No decrease in altitude detected in data file. \n");

    /*  Exit program.  */
    return EXIT_SUCCESS;
}
/*-----------------------------------------------------------------*/

/*-----------------------------------------------------------------*/
/*  Problem chapter3_14                                            */
/*                                                                 */
/*  This program reads velocity data from a data file and         */
/*  determines the number of stages in the rocket.  The file      */
/*  contains a trailer line with -99 for all values.              */

#include <stdio.h>
#include <stdlib.h>
#define INCREASE 1
#define DECREASE -1
#define FILENAME "rocket2.dat"

main()
{
    /*  Declare and initialize variables.  */
    int dir=INCREASE, stages=1;
    double time, altitude, acceleration, velocity=0.0, prev_vel=0;
    FILE *rocket;

    /*  Open input file.  */
    rocket = fopen(FILENAME,"r");

    /*  Read first set of data.  */
    fscanf(rocket,"%lf %lf %lf %lf",
           &time,&altitude,&velocity,&acceleration);
    prev_vel = velocity;

    /* Keep looking for the trailer while processing all the data */
    while (time > 0)
    {
```

```
        /*  Check for a change in direction.  */
        if ((prev_vel>velocity) && (dir==INCREASE))
           dir = DECREASE;
        else
           /*  New stage fired if velocity is increasing again.  */
           if ((prev_vel<velocity) && (dir==DECREASE))
           {
               stages++;
               dir = INCREASE;
           }

        /* Save values for comparison next time through loop */
        prev_vel=velocity;

        /* Read next data set */
        fscanf(rocket,"%lf %lf %lf %lf",
               &time,&altitude,&velocity,&acceleration);
    }

    /* Print results */
    printf("The number of stages was: %i\n",stages);
    return EXIT_SUCCESS;
}
/*------------------------------------------------------------------*/

/*------------------------------------------------------------------*/
/*  Problem chapter3_15                                             */
/*                                                                  */
/*  This program reads velocity data from a data file and           */
/*  determines the number of stages in the rocket.  The file        */
/*  contains a trailer line with -99 for all values.  The firing    */
/*  times for the rocket stages are printed.                        */

#include <stdio.h>
#include <stdlib.h>
#define INCREASE 1
#define DECREASE -1
#define FILENAME "rocket2.dat"

main()
{
    /* Declare variables */
    int dir=INCREASE, stages=1;
    double time, altitude, acceleration, velocity=0.0, prev_vel=0,
           fire_time;
    FILE *rocket;

    /*  Open input file.  */
    rocket = fopen(FILENAME,"r");

    /*  Read first set of data.  */
    fscanf(rocket,"%lf %lf %lf %lf",
           &time,&altitude,&velocity,&acceleration);
    prev_vel = velocity;
    fire_time = time;

    /*  Keep looking for the trailer while processing all the data.  */
    while (time > 0)
    {
        /*  Check for a change in direction. */
```

```
        if ((prev_vel>velocity) && (dir==INCREASE))
           dir = DECREASE;
        else
           /*  New stage fired if velocity is increasing again.  */
           if ((prev_vel<velocity) && (dir==DECREASE))
           {
              stages++;
              printf("Stage fired at time: %4f.\n",fire_time);
              dir = INCREASE;
           }

        /*  Save values for comparison next time through loop.  */
        fire_time = time;
        prev_vel = velocity;

        /*  Read next data set.  */
        fscanf(rocket,"%lf %lf %lf %lf",
              &time,&altitude,&velocity,&acceleration);
     }

     /*  Print results.  */
     printf("The number of stages was: %i\n",stages);

     /*  Exit program.  */
     return EXIT_SUCCESS;
}
/*-------------------------------------------------------------------*/

/*-------------------------------------------------------------------*/
/*  Problem chapter3_16                                              */
/*                                                                   */
/*  This program reads velocity data from a data file and           */
/*  determines the times during which the acceleration is due       */
/*  only to gravity.  The file does not contain a header or         */
/*  trailer line.                                                    */

#include <stdio.h>
#include <stdlib.h>
#define MAX_GRAVITY -9.31     /*  within -5% of gravity  */
#define MIN_GRAVITY  -10.29   /*  within  5% of gravity  */
#define GRAVITY -9.8
#define FILENAME "rocket3.dat"

main()
{
     /*  Declare variables.  */
     double time, altitude, acceleration, velocity;
     FILE *rocket;

     /*  Open input file for reading.  */
     rocket = fopen(FILENAME,"r");

     /*  Print times when acceleration is due only to gravity.  */
     while (fscanf(rocket,"%lf %lf %lf %lf",
           &time,&altitude,&velocity,&acceleration) == 4)
     {
        if ((acceleration<=MAX_GRAVITY) && (acceleration>=MIN_GRAVITY))
           printf("Acceleration due to gravity only at time: %.3f\n",
                 time);
     }
```

```
    /*  Exit program.  */
    return EXIT_SUCCESS;
}
/*-------------------------------------------------------------------*/

/*-------------------------------------------------------------------*/
/*  Problem chapter3_17                                          */
/*                                                               */
/*  This program reads a data file named suture.dat that contains */
/*  information on batches of sutures that have been rejected during */
/*  a 1-week period.  This program generates a report that computes */
/*  the percent of the batches rejected due to temperature, the  */
/*  percent rejected due to pressure, and the percent rejected due */
/*  to dwell time.                                               */

#include <stdio.h>
#include <stdlib.h>
#define MIN_TEMP 150
#define MAX_TEMP 170
#define MIN_PRESS 60
#define MAX_PRESS 70
#defile MIN_DWELL 2
#define MAX_DWELL 2.5
#define FILENAME "suture.dat"

main()
{
    /*  Define and initialize variables.  */
    int temp_rejects=0, press_rejects=0, dwell_rejects=0,
        batches_rejected=0, batch;
    double temp_per, press_per, dwell_per, temperature, pressure,
         dwell_time;
    FILE *suture;

    /*  Open input file.  */
    suture = fopen(FILENAME,"r");

    /*  Categorize rejected batches by failures       */
    /*   and record total number of rejected batches,  */
    /*   until there is no more data to process.       */
    while(fscanf(suture,"%i %lf %lf %lf",
        &batch,&temperature,&pressure,&dwell_time) == 4)
    {
        if ((temperature<MIN_TEMP) || (temperature>MAX_TEMP))
            temp_rejects++;
        if ((pressure<MIN_PRESS ) || (pressure>MAX_PRESS ))
            press_rejects++;
        if ((dwell_time<MIN_DWELL) || (dwell_time>MAX_DWELL))
            dwell_rejects++;
        batches_rejected++;
    }

    /*  Now calculate the percentages required and print them.  */
    temp_per = (double)temp_rejects /(double)batches_rejected;
    press_per = (double)press_rejects/(double)batches_rejected;
    dwell_per = (double)dwell_rejects/(double)batches_rejected;
    printf("Category        Percent rejected\n");
    printf("-----------------------------\n");
    printf("temperature        %f\n",temp_per);
```

```
      printf("pressure             %f\n",press_per);
      printf("dwell time           %f\n",dwell_per);
      printf("\n\nTotal number of batches: %i",batches_rejected);

      /*  Exit program.  */
      return EXIT_SUCCESS;
}
/*------------------------------------------------------------------*/

/*------------------------------------------------------------------*/
/*   Problem chapter3_18                                            */
/*                                                                  */
/*   This program modifies the solution to problem 17 such          */
/*   that it also prints the number of batches in each rejection    */
/*   category and the total number of batches rejected.             */

#include <stdio.h>
#include <stdlib.h>
#define MIN_TEMP 150
#define MAX_TEMP 170
#define MIN_PRESS 60
#define MAX_PRESS 70
#define MIN_DWELL 2
#define MAX_DWELL 2.5
#define FILENAME "suture.dat"

main()
{
   /*  Define and initialize variables.  */
   int temp_rejects=0, press_rejects=0, dwell_rejects=0,
       batches_rejected=0, batch;
   double temp_per, press_per, dwell_per, temperature, pressure,
          dwell_time;
   FILE *suture;

   /*  Open data file.  */
   suture = fopen(FILENAME,"r");

   /*  Categorize rejected batches by failures      */
   /*  and record total number of rejected batches, */
   /*  until there is no more data to process.      */
   while(fscanf(suture,"%i %lf %lf %lf",&batch,
          &temperature,&pressure,&dwell_time) == 4)
   {
      if ((temperature<MIN_TEMP) || (temperature>MAX_TEMP))
         temp_rejects++;
      if ((pressure<MIN_PRESS ) || (pressure>MAX_PRESS ))
         press_rejects++;
      if ((dwell_time<MIN_DWELL) || (dwell_time>MAX_DWELL))
         dwell_rejects++;
      batches_rejected++;
   }

   /*  Now calculate the percentages required and print them.  */
   temp_per = (double)temp_rejects /(double)batches_rejected;
   press_per = (double)press_rejects/(double)batches_rejected;
   dwell_per = (double)dwell_rejects/(double)batches_rejected;
   printf("Category         Percent rejected\n");
   printf("--------------------------------\n");
   printf("temperature          %f\n",temp_per);
```

24

```
      printf("pressure            %f\n",press_per);
      printf("dwell time          %f\n",dwell_per);

      printf("\n\nCategory      Number of batches rejected\n");
      printf("----------------------------------------\n");
      printf("temperature         %i\n",temp_rejects);
      printf("pressure            %i\n",press_rejects);
      printf("dwell time          %i\n",dwell_rejects);
      printf("\n\nTotal number of batches: %i\n\n",batches_rejected);

      /*  Exit program.  */
      return EXIT_SUCCESS;
}
/*------------------------------------------------------------------*/

/*------------------------------------------------------------------*/
/*  Problem chapter3_19                                             */
/*                                                                  */
/*  This program reads the data file suture.dat and makes           */
/*  sure that the information relates only to batches that should    */
/*  have been rejected.  If any batch should not be in the data      */
/*  file, an error message with the batch information is printed.    */

#include <stdio.h>
#include <stdlib.h>
#define MIN_TEMP 150
#define MAX_TEMP 170
#define MIN_PRESS 60
#define MAX_PRESS 70
#define MIN_DWELL 2
#define MAX_DWELL 2.5
#define TRUE 1
#define FALSE 0
#define FILENAME "suture.dat"

main()
{
   /*  Define and initialize variables.  */
   int reject, batch;
   double temperature, pressure, dwell_time;
   FILE *suture;

   /*  Open input file.  */
   suture = fopen(FILENAME,"r");

   /*  Check that the data really is for rejected suture packages.  */
   while(fscanf(suture,"%i %lf %lf %lf",&batch,&temperature,
         &pressure,&dwell_time) == 4)
   {
      reject = FALSE;
      if ((temperature<MIN_TEMP) || (temperature>MAX_TEMP))
         reject = TRUE;
      if ((pressure<MIN_PRESS ) || (pressure>MAX_PRESS ))
         reject = TRUE;
      if ((dwell_time<MIN_DWELL) || (dwell_time>MAX_DWELL))
         reject = TRUE;
      if (reject == FALSE)
      {
         printf("Batch %i should not be rejected.\n",batch);
         printf("   Temperature: %f\n",temperature);
```

25

```
            printf("   Pressure:       %f\n",pressure);
            printf("   Dwell Time:    %f\n\n",dwell_time);
        }
    }

    /*  Exit program.  */
    return EXIT_SUCCESS;
}
/*-------------------------------------------------------------------*/


/*-------------------------------------------------------------------*/
/*  Problem chapter3_20                                              */
/*                                                                   */
/*  Assume that there are 14,000 acres total with 2500 acres         */
/*  uncut, and that the reforestation rate is 0.02.                  */
/*  This program prints a table showing the number of acres          */
/*  forested at the end of each year, for a total of 20 years.       */

#include <stdio.h>
#include <stdlib.h>
#define REFOREST_RATE 0.02
#define UNCUT_ACRES 2500
#define MAX_YEARS 20

main()
{
    /*  Declare and initialize variables.  */
    int year=1;
    double forested=UNCUT_ACRES;

    /*  Print headings.  */
    printf("YEAR     FORESTED ACRES AT END OF YEAR\n");
    printf("----------------------------------\n");

    /*  Print amount forsted for MAX_YEARS.  */
    while (year <= MAX_YEARS)
    {
        forested += REFOREST_RATE*forested;
        printf("%3i          %f\n",year,forested);
        year++;
    }

    /*  Exit program.  */
    return EXIT_SUCCESS;
}
/*-------------------------------------------------------------------*/


/*-------------------------------------------------------------------*/
/*  Problem chapter3_21                                              */
/*                                                                   */
/*  This program modifies the solution to problem 20 so that         */
/*  the user can enter the numvber of years for the table.           */

#include <stdio.h>
#include <stdlib.h>
#define REFOREST_RATE 0.02
#define UNCUT_ACRES 2500

main()
```

```c
{
    /*  Declare and initialize variables.  */
    int year=1, max_years=0;
    double forested=UNCUT_ACRES;

    /*  Print current information.  */
    printf("Uncut acres: %i \n" ,UNCUT_ACRES);
    printf("Reforestation rate: %f \n\n", REFOREST_RATE);

    /*  Enter number of years for table.  */
    while (max_years < 1)
    {
        printf("Enter number of years for reforestation calculations:");
        scanf("%i", &max_years);
    }

    /*  Print headings.  */
    printf("\n\nYEAR     FORESTED ACRES AT END OF YEAR\n");
    printf("-----------------------------------\n");

    /*  Compute and print reforested acres.  */
    while (year <= max_years)
    {
        forested += REFOREST_RATE*forested;
        printf("%3i          %f\n", year,forested);
        year++;
    }

    /*  Exit program.  */
    return EXIT_SUCCESS;
}
/*------------------------------------------------------------------*/

/*------------------------------------------------------------------*/
/*  Problem chapter3_22                                             */
/*                                                                  */
/*  This program modifies the solution to problem 20 so that the    */
/*  user can enter the number of acres and the program will         */
/*  determine how many years are required for the number            */
/*  of acres to be completely reforested.                           */

#include <stdio.h>
#include <stdlib.h>
#define REFOREST_RATE 0.02
#define UNCUT_ACRES 2500
#define MAX_ACRES 14000

main()
{
    /*  Define and initialize variables.  */
    int year=0;
    double forested=UNCUT_ACRES, acres=UNCUT_ACRES-1;

    /*  Print headers.  */
    printf("MAXIMUM number of acres to reforest: %i \n",MAX_ACRES);
    printf("MINIMUM number of acres: %i \n\n",UNCUT_ACRES);

    /*  Read acres to be reforested.  */
    while ((acres>MAX_ACRES) || (acres<UNCUT_ACRES))
    {
```

```
      printf("Enter acres to be reforested: ");
      scanf("%lf", &acres);
   }

   /*   Print headings.   */
   printf("\n\nYEAR      FORESTED ACRES AT END OF YEAR \n");
   printf("------------------------------------ \n");

   /*   Compute and print yearly reforestation information.   */
   while (forested <= acres)
   {
      year++;
      forested += REFOREST_RATE*forested;
      printf("%3i            %7.2f\n", year,forested);
   }

   /*   Print years required for reforestation.   */
   printf("\n\nUp to %3i years are required to reforest %7.2f"
          " acres.\n",year,acres);

   /*   Exit program.   */
   return EXIT_SUCCESS;
}
/*------------------------------------------------------------------*/


/*------------------------------------------------------------------*/
/*   Problem chapter3_23                                            */
/*                                                                  */
/*   This program determines the number of days that had            */
/*   temperatures in the following categories for January 1991:     */
/*      Below 0, 0-32, 33-50, 51-60, 61-70, over 70                 */

#include <stdio.h>
#include <stdlib.h>
#define DAYS_IN_MONTH 31
#define DATA_PER_DAY 32
#define FILENAME "jan91.dat"
main()
{
   /*   Declare and initialize variables.   */
   int verycold=0, cold=0, chilly=0, mild=0, warm=0, verywarm=0,
       days, i;
   double min_temp, max_temp, dummy;
   FILE *weather;

   /*   Open input file.   */
   weather = fopen(FILENAME,"r");

   /*   Read data and add to correct categories.   */
   for (days=0; days<DAYS_IN_MONTH; days++)
   {
      fscanf(weather,"%lf %lf %lf",&dummy,&max_temp,&min_temp);

      /*   Figure out the temperature ranges during the day.   */
      verycold++;
      if (min_temp < 0)
         verycold++;
      if ((min_temp<=32) && (max_temp>=0))
         cold++;
      if ((min_temp<=50) && (max_temp>=33))
```

```
            chilly++;
         if ((min_temp<=60) && (max_temp>=51))
            mild++
         if ((min_temp<=70) && (min_temp>=61))
            warm++
         if (max_temp>70)
            verywarm++

         /*  Ignore the rest of the data on this line.  */
         for (i=3; i<DATA_PER_DAY; i++)
            fscanf(weather,"%lf",&dummy);
      }

   /*  Print results.  */
   printf("Number of days in each category in January 1991\n\n");
   printf("Below zero\t0-32\t33-50\t51-60\t61-70\tOver 70\n");
   printf("%i\t\t%i\t%i\t%i\t%i\t%i\n",
          verycold,cold,chilly,mild,warm,verywarm);

   /*  Exit program.  */
   return EXIT_SUCCESS;
}
/*------------------------------------------------------------------*/


/*------------------------------------------------------------------*/
/*  Problem chapter3_24                                             */
/*                                                                  */
/*  This program determines the percentage of days that had        */
/*  temperatures in the following categories for January 1991:     */
/*     Below 0, 0-32, 33-50, 51-60, 61-70, over 70                 */

#include <stdio.h>
#include <stdlib.h>
#define DAYS_IN_MONTH 31
#define DATA_PER_DAY 32
#define FILENAME "jan91.dat"
main()
{
   /*  Declare variables.  */
   int verycold=0, cold=0, chilly=0, mild=0, warm=0, verywarm=0,
       days, i;
   double min_temp, max_temp, dummy, convert=100.0/DAYS_IN_MONTH;
   FILE *weather;

   /*  Open input file.  */
   weather = fopen(FILENAME,"r");

   /*  Read data and add to correct categories.  *.
   for (days=0; days<DAYS_IN_MONTH; days++)
   {
      fscanf(weather,"%lf %lf %lf",&dummy,&max_temp,&min_temp);

      /* Figure out the temperature ranges during the day */
      verycold++;
      if (min_temp < 0)
         verycold++;
      if ((min_temp<=32) && (max_temp>=0))
         cold++;
      if ((min_temp<=50) && (max_temp>=33))
         chilly++;
```

29

```
            if ((min_temp<=60) && (max_temp>=51))
                mild++
            if ((min_temp<=70) && (min_temp>=61))
                warm++
            if (max_temp>70)
                verywarm++

            /* Ignore the rest of the data on this line */
            for (i=3; i<DATA_PER_DAY; i++)
                fscanf(weather,"%lf",&dummy);
        }

        /*  Print results.  */
        printf("Percentage of days in each category in January 1991\n\n");
        printf("Below zero\t0-32\t33-50\t51-60\t61-70\tOver 70\n");
        printf("%.2f\t\t%.2f\t%.2f\t%.2f\t%.2f\t%.2f\n",convert*verycold,
                convert*cold,convert*chilly,convert*mild,convert*warm,
                convert*verywarm);

        /*  Exit program.  */
        return EXIT_SUCCESS;
}
/*-------------------------------------------------------------------*/


/*-------------------------------------------------------------------*/
/*  Problem chapter3_25                                              */
/*                                                                   */
/*  This program determines the number of days that had             */
/*  temperatures in the following categories for May-Aug 1991:       */
/*     Below 0, 0-32, 33-50, 51-60, 61-70, over 70                  */

#include <stdio.h>
#include <stdlib.h>
#define VERYCOLD 0
#define COLD 33
#define CHILLY 51
#define MILD 61
#define WARM 71
#define DATA_PER_DAY 32
#define NUM_MONTHS 4
#define FILE1 "may91.dat"
#define FILE2 "jun91.dat"
#define FILE3 "jul91.dat"
#define FILE4 "aug91.dat"

main()
{
    /*  Declare and initialize variables.  */
    int verycold=0, cold=0, chilly=0, mild=0, warm=0, verywarm=0,
        i, months;
    double min_temp, max_temp, dummy;
    FILE *weather;

    /*  Print title and headers.  */
    printf("Number of days in each category for May to August 1991\n\n");
    printf("Month\tBelow zero\t0-32\t33-50\t51-60\t61-70\tOver 70\n");

    /*  Process each month of data.  */
    for (months=0; months<NUM_MONTHS; months++)
    {
```

```
        /*  Open correct input file.  */
        switch(months)
        {
            case 0:
                weather = fopen(FILE1,"r");
                printf("May\t");
                break;
            case 1:
                weather = fopen(FILE2,"r");
                printf("June\t");
                break;
            case 2:
                weather = fopen(FILE3,"r");
                printf("July\t");
                break;
            case 3:
                weather = fopen(FILE4,"r");
                printf("August\t");
                break;
            default:
                return EXIT_FAILURE;
        }

        /*  Read data and add to correct categories.  */
        while (fscanf(weather,"%lf %lf %lf",
                &dummy,&max_temp,&min_temp) ==3)
        {
            /*  Figure out the temperature ranges during the day.  */
            if (min_temp < 0)
                verycold++;
            if ((min_temp<=32) && (max_temp>=0))
                cold++;
            if ((min_temp<=50) && (max_temp>=33))
                chilly++;
            if ((min_temp<=60) && (max_temp>=51))
                mild++
            if ((min_temp<=70) && (min_temp>=61))
                warm++
            if (max_temp>70)
                verywarm++

            /* Ignore the rest of the data on this line */
            for (i=3; i<DATA_PER_DAY; i++)
                fscanf(weather,"%lf",&dummy);
        }
    }

    /*  Print results.  */
    printf("%i\t\t%i\t%i\t%i\t%i\t%i\n",
            verycold,cold,chilly,mild,warm,verywarm);

    /*  Exit program.  */
    return EXIT_SUCCESS;
}
/*------------------------------------------------------------------*/

/*------------------------------------------------------------------*/
/*  Problem chapter3_26                                             */
/*                                                                  */
/*  This program determines the average temperature for            */
```

```c
/*  days with fog in November 1991.                                 */

#include <stdio.h>
#include <stdlib.h>
#define DATA_PER_DAY 32
#define FILE1 "nov91.dat"
#define FOG_DATA 8
#define FOG 1

main()
{
   /*  Declare and initialize variables.   */
   int i, types, foggy=0, ave_foggy=0.0;
   double ave_temp, dummy;
   FILE *weather;

   /*  Open input file.   */
   weather = fopen(FILE1,"r");

   /*  Read data and count foggy days.   *
   while (fscanf(weather,"%lf %lf %lf %lf",
         &dummy,&dummy,&dummy,&ave_temp) ==4)
   {
      /*  Ignore data up to the fog information.   */
      for (i=4; i<FOG_DATA; i++)
         fscanf(weather,"%lf",&dummy);

      /*  Read this column for fog information.   */
      fscanf(weather,"%i",&types);
      if (types == FOG)
      {
         foggy++;
         ave_foggy += ave_temp;
      }

      /*  Skip the rest of the line of data.   */
      for (i=FOG_DATA+1; i<DATA_PER_DAY; i++)
         fscanf(weather,"%lf",&dummy);
   }

   /*  Print results.   */
   if (foggy >0 )
      printf("Average temperature for the %i foggy days in "
             "Nov 1991: %.3f\n",foggy,ave_foggy/foggy);
   else
      printf("No foggy days in Nov 1991.\n");

   /*  Close file and exit program.   */
   fclose(weather);
   return EXIT_SUCCESS;
}
/*-------------------------------------------------------------------*/


/*-------------------------------------------------------------------*/
/*  Problem chapter3_27                                              */
/*                                                                   */
/*  This program determines the date in December 1991 with the       */
/*  largest difference between the maximum temperature and the        */
/*  minimum temperature.  The date, the temperatures, and the         */
/*  differences are printed.                                          */
```

```c
#include <stdio.h>
#include <stdlib.h>
#define DATA_PER_DAY 32
#define FILE1 "dec91.dat"

main()
{
   /*  Declare and initialize variables.  */
   int dayofmonth, day, i;
   double max_temp, min_temp, dummy, difference, big_difference=0,
          result_max, result_min;
   FILE *weather;

   /*  Open input file.  */
   weather = fopen(FILE1,"r");

   /*  Read data and find largest difference.  */
   while (fscanf(weather,"%i %lf %lf",&day,&max_temp,&min_temp) ==3)
   {
      difference = max_temp - min_temp;
      if (big_difference < difference)
      {
         dayofmonth = day;
         big_difference = difference;
         result_max = max_temp;
         result_min = min_temp;
      }

      /*  Skip the rest of the line of data.  */
      for (i=3; i<DATA_PER_DAY; i++)
         fscanf(weather,"%lf",&dummy);
   }

   /*  Print results.  */
   printf("The maximum difference in temperatures was: %.3f "
          "Farenheit degrees.\n",big_difference);
   printf("It occured on December %i. \n",dayofmonth);
   printf("Max temperature was: %.3f, and the min temperature "
          "was %.3f\n",result_max,result_min);

   /*  Close file and exit program.   */
   fclose(weather);
   return EXIT_SUCCESS;
}
/*------------------------------------------------------------------*/

/*------------------------------------------------------------------*/
/*   Problem chapter3_28                                            */
/*                                                                  */
/*   This program reads the critical path information               */
/*   and prints a project completion timetable that lists           */
/*   each even number, the maximum number of days for a             */
/*   task within the even, and the total number of days for         */
/*   the project completion.                                        */

#include <stdio.h>
#include <stdlib.h>
#define FILENAME "path.dat"
```

```
main()
{
    /*  Declare and initialize variables.  */
    int event, previous_event=1, task=0, days=0, max_days=0,
        project_days=0;
    FILE *path;

    /*  Open data file.  */
    path=fopen(FILENAME,"r");

    /*  Set up the headers before the loop.  */
    printf("Event      Max Number Days for a Task \n");
    printf("--------------------------------- \n");
    while(fscanf(path,"%i %i %i",&event,&task,&days) == 3)
    {
        if (event == previous_event)
        {
            /*  Count tasks for the current event.  */
            if (days>max_days)
                max_days = days;
        }
        else
        {   /*  A new event, so we can finish up the current one  */
            /*   and start keeping values for the new one.         */
            project_days += max_days;
            printf("%i          %i \n",previous_event,max_days);
            previous_event = event;
            max_days = days;
        }
    }

    /*  Count the last event, and print the total project duration.  */
    project_days += max_days;
    printf("%i          %i \n",event,max_days);
    printf("\nProject duration: %i \n",project_days);

    /*  Close file and exit program.  */
    fclose(path);
    return EXIT_SUCCESS;
}
/*-------------------------------------------------------------------*/


/*-------------------------------------------------------------------*/
/*  Problem chapter3_29                                              */
/*                                                                   */
/*  This program reads the critical path information                 */
/*  and print a report that lists the event number and              */
/*  task number for all tasks requiring more than 5 days.           */

#include <stdio.h>
#include <stdlib.h>
#define FILENAME "path.dat"
#define MAX_DAYS 5

main()
{
    /*  Declare variables.  */
    int event, task, days;
    FILE *path;
```

```
    /*  Open input file.  */
    path = fopen(FILENAME,"r");

    /*  Set up the headers before the loop.  */
    printf("Tasks requiring more than %i days:\n\n",MAX_DAYS);
    printf("Event     Task     Days \n");
    printf("---------------------- \n");

    /*  Print tasks requiring more than MAX_DAYS days.  */
    while(fscanf(path,"%i %i %i",&event,&task,&days) == 3)
  {
      if (days > MAX_DAYS)
         printf("%3i    %3i    %3i \n",event,task,days);
  }

    /*  Close file and exit program.  */
    fclose(path);
    return EXIT_SUCCESS;
}
/*-------------------------------------------------------------------*/

/*-------------------------------------------------------------------*/
/*  Problem chapter3_30                                              */
/*                                                                   */
/*  This program reads the critical path information                 */
/*  and prints a report that lists the number of each event          */
/*  and a count of the number of tasks within the event.             */

#include <stdio.h>
#include <stdlib.h>
#define FILENAME "path.dat"

main()
{
    /*  Declare and initialize variables.  */
    int event, previous_event=1, task, days, number_tasks=0;;
    FILE *path;

    /*  Open input file.  */
    path=fopen(FILENAME,"r");

    /*  Set up the headers before the loop   */
    printf("Event     Number of Tasks \n");
    printf("--------------------------------- \n");

    /*  Read data and print event information.  */
    while(fscanf(path,"%i %i %i",&event,&task,&days) == 3)
    {
       if (event == previous_event)
          /*  Count tasks for the current event.  */
          number_tasks++;
       else
       { /*  A new event, so we can finish up the one and  */
         /*   start keeping values for the new one.         */
         printf("%i          %i \n",previous_event,number_tasks);
         previous_event = event;
         number_tasks = 1;
       }
    }
```

```c
    /*  Printthe last event.  */
    printf("%i           %i \n",event,number_tasks);

    /*  Close file and exit program.  */
    fclose(path);
    return EXIT_SUCCESS;
}
/*-------------------------------------------------------------------*/
```

Chapter 4

```
/*------------------------------------------------------------------*/
/*  Problem chapter4_1                                              */
/*                                                                  */
/*  This program simulates tossing a "fair" coin.                   */
/*  The user enters the number of tosses.                          */

#include <stdio.h>
#include <stdlib.h>
#define HEADS 1
#define TAILS 0

main()
{
   /*  Declare and initialize variables  */
   /*  and declare function prototypes.  */
   int tosses=0, heads=0, required=0;
   int rand_int(int a, int b);

   /*  Prompt user for number of tosses.  */
   printf("\n\nEnter number of fair coin tosses: ");
   scanf("%i",&required);
   while (required <= 0)
   {
      printf("Tosses must be an integer number, greater than zero.\n\n");
      printf("Enter number of fair coin tosses: ");
      scanf("%i",&required);
    }

   /*  Toss coin the required number of times, and keep track of  */
   /*  the number of heads.  Use rand_int for the "toss" and      */
   /*  and consider a positive number to be "heads."              */
   while (tosses < required)
   {
      tosses++;
      if (rand_int(TAILS,HEADS) == HEADS)
                heads++;
   }

   /*  Print results.  */
   printf("\n\nNumber of tosses: %i\n",tosses);
   printf("Number of heads:    %i\n",heads);
   printf("Number of tails:    %i\n",tosses-heads);
   printf("Percentage of heads: %f\n",100.0 * heads/tosses);
   printf("Percentage of tails: %f\n",100.0 * (tosses-heads)/tosses);

   /*  Exit program.  */
   return EXIT_SUCCESS;
}
/*------------------------------------------------------------------*/
          (rand_int function from page 168)
/*------------------------------------------------------------------*/


/*------------------------------------------------------------------*/
/*  Problem chapter4_2                                              */
/*                                                                  */
/*  This program simulates tossing an "unfair" coin.               */
/*  The user enters the number of tosses.                          */
```

```
#include <stdio.h>
#include <stdlib.h>
#define   HEADS   10
#define   TAILS   1
#define   WEIGHT  6
main()
{
    /*  Declare variables and function prototypes.  */
    int tosses=0, heads=0, required=0;
    int rand_int(int a, int b);

    /*  Prompt user for number of tosses.  */
    printf("\n\nEnter number of unfair coin tosses: ");
    scanf("%i",&required);

    /*  Toss coin the required number of times, and  */
    /*   keep track of the number of heads.          */
    while (tosses < required)
    {
        tosses++;
        if (rand_int(TAILS,HEADS) <= WEIGIIT)
            heads++;
    }

    /*  Print results.  */
    printf("\n\nNumber of tosses:     %i\n",tosses);
    printf("Number of heads:      %i\n",heads);
    printf("Number of tails:      %i\n",tosses-heads);
    printf("Percentage of heads:  %f\n", 100.0 * heads/tosses);
    printf("Percentage of tails:  %f\n", 100.0 * (tosses-heads)/tosses);

    /*  Exit program.  */
    return EXIT_SUCCESS;
}
/*------------------------------------------------------------------*/
              (rand_int function from page 168)
/*------------------------------------------------------------------*/


/*------------------------------------------------------------------*/
/*   Problem chapter4_3                                           */
/*                                                                */
/*   This program simulates rolling a six-sided "fair" die.       */
/*   The user enter the number of rolls.                          */

#include <stdio.h>
#include <stdlib.h>
#define MIN 1
#define MAX 6

main()
{
    /* Declare variables and function prototypes.  */
    int onedot=0, twodots=0, threedots=0, fourdots=0, fivedots=0,
        sixdots=0, rolls=0, required=0;
    int rand_int(int a, int b);

    /*  Prompt user for number of rolls.  */
    printf("\n\nEnter number of fair die rolls: ");
    scanf("%i", &required);
```

38

```c
    /*  Roll the die as many times as required  */
    while (rolls < required)
    {
        rolls++;
        switch(rand_int(MIN,MAX))
        {
            case 1: onedot++;
                    break;
            case 2: twodots++;
                    break;
            case 3: threedots++;
                    break;
            case 4: fourdots++;
                    break;
            case 5: fivedots++;
                    break;
            case 6: sixdots++;
                    break;
            default:
                    printf("\nDie roll result out of range!\n");
                    return EXIT_FAILURE;
                    break;
        }
    }

    /*  Print the results.  */
    printf("\nNumber of rolls:  %i\n",rolls);
    printf("Number of ones:   %i, percentage: %f\n",
            onedot,100.0*onedot/rolls);
    printf("Number of twos:   %i, percentage: %f\n",
            twodots,100.0*twodots/rolls);
    printf("Number of threes: %i, percentage: %f\n",
            threedots,100.0*threedots/rolls);
    printf("Number of fours:  %i, percentage: %f\n",
            fourdots,100.0*fourdots/rolls);
    printf("Number of fives:  %i, percentage: %f\n",
            fivedots,100.0*fivedots/rolls);
    printf("Number of sixes:  %i, percentage: %f\n",
            sixdots,100.0*sixdots/rolls);

    /*  Exit program.  */
    return EXIT_SUCCESS;
}
/*-------------------------------------------------------------------*/
            (rand_int function from page 168)
/*-------------------------------------------------------------------*/

/*-------------------------------------------------------------------*/
/*  Problem chapter4_4                                               */
/*                                                                   */
/*  This program simulates an experiment rolling two six-sided       */
/*  "fair" dice.  The user enters the number of rolls.               */

#include <stdio.h>
#include <stdlib.h>
#define MIN 1
#define MAX 6
#define TOTAL 8
```

```
main()
{
    /*  Declare variables and function prototypes.   */
    int rolls=0, die1, die2, required=0, sum=0;
    int rand_int(int a, int b);

    /*  Prompt user for number of rolls.  */
    printf("\n\nEnter number of fair dice rolls: ");
    scanf("%u",&required);

    /*  Roll the die as many times as required.  */
    while (rolls < required)
    {
        rolls++;
        die1=rand_int(MIN,MAX);
        die2=rand_int(MIN,MAX);
        if (die1+die2 == TOTAL)
            sum++;
        printf("Results: %u %u\n",die1,die2);
    }

    /*  Print the results.  */
    printf("\n\nNumber of rolls:  %i\n",rolls);
    printf("Number of %is:      %u\n",TOTAL,sum);
    printf("Percentage of %is: %f\n",TOTAL,100.0*sum/rolls);

    /*  Exit program.  */
    return EXIT_SUCCESS;
}
/*------------------------------------------------------------------*/
            (rand_int function from page 168)
/*------------------------------------------------------------------*/

/*------------------------------------------------------------------*/
/*  Problem chapter4_5                                              */
/*                                                                  */
/*  This program simulates a lottery drawing that uses balls        */
/*  numbered from 1 to 10.                                          */

#include <stdio.h>
#include <stdlib.h>
#define MIN 1
#define MAX 10
#define NUMBER 7
#define NUM_BALLS 3

main()
{
    /*  Define variables and function prototypes.   */
    unsigned int alleven = 0, num_in_sim = 0, onetwothree = 0, first,
                second, third;
    int lotteries = 0, required = 0;
    int rand_int(int a, ,int b);

    /*  Prompt user for the number of lotteries.  */
    printf("\n\nEnter number of lotteries: ");
    scanf("%i",&required);
    while (required <= 0)
    {
        printf("The number of lotteries must be an integer number, "
```

40

```c
                      "greater than zero.\n\n");
        printf("Enter number of lotteries: ");
        scanf("%i",&required);
    }

    /*  Get three lottery balls, check for even or odd, and NUMBER.  */
    /*  Also check for the 1-2-3 sequence and its permutations.      */
    while (lotteries < required)
    {
        lotteries++;

         /*  Draw three unique balls.  */
        first = rand_int(MIN,MAX);
        do
            second = rand_int(MIN,MAX);
        while (second == first);
        do
            third = rand_int(MIN,MAX);
        while ((second==third) || (first==third));

        printf("Lottery number is: %u-%u-%u\n",first,second,third);

        /*  Are they all even?  */
        if ((first % 2 == 0) && (second % 2 == 0) && (third %2 == 0))
            alleven++;

        /*  Are any of them equal to NUMBER?  */
        if ((first == NUMBER) || (second==NUMBER) || (third == NUMBER))
            num_in_sim++;

        /*  Are they 1-2-3 in any order?  */
        if ((first <= 3) && (second <= 3) && (third <= 3))
            if ((first != second) && (first != third) && (second != third))
                onetwothree++;
    }

    /* Print results.  */
    printf("\nPercentage of time the result contains three even numbers:"
           " %f\n",100.0*alleven/lotteries);
    printf("Percentage of time the number %u occurs in the three"
           " numbers:  %f\n",NUMBER,100.0*num_in_sim/lotteries);
    printf("Percentage of time the numbers 1,2,3 occur (not necessarily"
           " in order): %f\n",100.0*onetwothree/lotteries);

    /*  Exit program.  */
    return EXIT_SUCCESS;

}
/*-------------------------------------------------------------------------*/
/*          (rand_int function from page 168)                              */
/*-------------------------------------------------------------------------*/


/*-------------------------------------------------------------------------*/
/*  Problem chapter4_6                                                      */
/*                                                                         */
/*                                                                         */
/*  This program simulates the design in Figure 4-10 using a               */
/*  component reliability of 0.8 for component 1, 0.85 for                  */
/*  component 2 and 0.95 for component 3.  The estimate of the             */
/*  reliability is computed using 5000 simulations.                        */
/*  (The analytical reliability of this system is 0.794.)                  */
```

```
#include <stdio.h>
#include <stdlib.h>
#define SIMULATIONS 5000
#define REL1 0.8
#define REL2 0.85
#define REL3 0.95
#define MIN_REL 0
#define MAX_REL 1

main()
{
    /*  Define variables and function prototypes.  */
    int num_sim=0, success=0;
    double est1, est2, est3;
    double rand_float(double a, double b);

    /*  Run simulations.  */
    for (num_sim=0; num_sim<SIMULATIONS; num_sim++)
    {
        /* Get the random numbers */
        est1 = rand_float(MIN_REL,MAX_REL);
        est2 = rand_float(MIN_REL,MAX_REL);
        est3 = rand_float(MIN_REL,MAX_REL);

        /* Now test the configuration */
        if ((est1<=REL1) && ((est2<=REL2) || (est3<=REL3)))
            success++;
    }

    /*  Print results.  */
    printf("Simulation Reliability for %i trials: %f\n",
            num_sim,(double)success/num_sim);

    /*  Exit program.  */
    return EXIT_SUCCESS;
}
/*------------------------------------------------------------------*/
            (rand_float function from page 170)
/*------------------------------------------------------------------*/


/*------------------------------------------------------------------*/
/*  Problem chapter4_7                                              */
/*                                                                  */
/*  This program simulates the design in Figure 4-11 using a        */
/*  component reliability of 0.8 for components 1 and 2,            */
/*  and 0.95 for components 3 and 4.  Print the estimate of the     */
/*  reliability using 5000 simulations.                            */
/*  (The analytical reliability of this system is 0.9649.)          */

#include <stdio.h>
#include <stdlib.h>
#define SIMULATIONS 5000
#define REL12 0.8
#define REL34 0.95
#define MIN_REL 0
#define MAX_REL 1

main()
{
```

```c
    /* Define variables and function prototypes.  */
    int num_sim=0, success=0;
    double est1, est2, est3, est4;
    double rand_float(double a, double b);

    /* Run simulations.  */
    for (num_sim=0; num_sim<SIMULATIONS; num_sim++)
    {
       /* Get the random numbers.  */
       est1 = rand_float(MIN_REL,MAX_REL);
       est2 = rand_float(MIN_REL,MAX_REL);
       est3 = rand_float(MIN_REL,MAX_REL);
       est4 = rand_float(MIN_REL,MAX_REL);

       /* Now test the configuration.  */
       if (((est1<=REL12) && (est2<=REL12)) ||
           ((est3<=REL34) && (est4<=REL34)))
          success++;
    }

    /* Print results.  */
    printf("Simulation Reliability for %i trials: %f\n",
           num_sim, (double)success/num_sim);

    /* Exit program.  */
    return EXIT_SUCCESS;
}
/*-------------------------------------------------------------------*/
            (rand_float function from page 170)
/*-------------------------------------------------------------------*/

/*-------------------------------------------------------------------*/
/*  Problem chapter4_8                                               */
/*                                                                   */
/*  This program simulates the design in Figure 4-12 using a        */
/*  component reliability of 0.95 for all components.  Print the     */
/*  estimate of the reliability using 5000 simulations.             */
/*  (The analytical reliability of this system is 0.99976.)         */

#include <stdio.h>
#include <stdlib.h>
#define SIMULATIONS 5000
#define REL 0.95
#define MIN_REL 0
#define MAX_REL 1

main()
{
    /* Define variables and function prototypes.  */
    int num_sim=0, success=0;
    double est1, est2, est3, est4;
    double rand_float(double a, double b);

    /* Run simulations.  */
    for (num_sim=0; num_sim<SIMULATIONS; num_sim++)
    {
       /* Get the random numbers.  */
       est1 = rand_float(MIN_REL,MAX_REL);
       est2 = rand_float(MIN_REL,MAX_REL);
       est3 = rand_float(MIN_REL,MAX_REL);
```

```
            est4 = rand_float(MIN_REL,MAX_REL);

            /*  Now test the configuration.  */
            if (((est1<=REL) || (est2<=REL)) || ((est3<=REL) && (est4<=REL)))
                success++;
        }

        /*  Print results.  */
        printf("Simulation Reliability for %i trials: %f\n",
            num_sim, (double)success/num_sim);

        /*  Exit program.  */
        return EXIT_SUCCESS;
}
/*----------------------------------------------------------------*/
/*           (rand_float function from page 170)                  */
/*----------------------------------------------------------------*/

/*----------------------------------------------------------------*/
/*  Problem chapter4_9                                          */
/*                                                              */
/*  This program generates a data file named wind.dat that      */
/*  contains one hour of simulated wind speeds.                 */

#include <stdio.h>
#include <stdlib.h>
#define DELTA_TIME 10
#define START_TIME 0
#define STOP_TIME 3600
#define FILENAME "wind.dat"

main()
{
    /*  Define variables and function prototypes.  */
    int timer=START_TIME;
    double ave_wind=0.0, gust_min=0.0, gust_max=0.0, windspeed=0.0;
    FILE *wind_data;
    double rand_float(double a, double b);

    /*  Open output file.  */
    wind_data = fopen(FILENAME,"w");

    /*  Prompt user for input and verify.  */
    printf("Enter average wind speed: ");
    scanf("%lf",&ave_wind);
    printf("Enter minimum gust: ");
    scanf("%lf",&gust_min);
    printf("Enter maximum gust (> minimum gust): ");
    scanf("%lf",&gust_max);

    for (timer=START_TIME; timer<=STOP_TIME; timer+=DELTA_TIME)
    {
        windspeed = rand_float(ave_wind+gust_min,ave_wind+gust_max);
        fprintf(wind_data,"%i %f\n",timer,windspeed);
    }

    /*  Close file and exit program.  */
    fclose(wind_data);
    return EXIT_SUCCESS;
}
```

```
/*------------------------------------------------------------------*/
            (rand_float function from page 170)
/*------------------------------------------------------------------*/


/*------------------------------------------------------------------*/
/*  Problem chapter4_10                                             */
/*                                                                  */
/*  This program generates flight simulator wind data with a        */
/*  0.5% possiblity of encountering a small storm at each time.     */

#include <stdio.h>
#include <stdlib.h>
#define DELTA_TIME 10
#define START_TIME 0
#define STOP_TIME 3600
#define FILENAME "wind.dat"
#define STORM_WIND 10
#define MIN_PROB 0
#define MAX_PROB 1
#define STORM_PROB 0.005
#define STORM 1
#define NO_STORM 0
#define MAX_DURATION 300

main()
{
    /*  Define variables and function prototypes.  */
    double ave_wind=0.0, gust_min=0.0, gust_max=0.0, windspeed = 0.0;
    int timer=START_TIME, storm_duration=0, storm_flag=NO_STORM;
    FILE *wind_data;
    double rand_float(double a, double b);

    /* Open file for writing */
    wind_data = fopen(FILENAME,"w");

    /* Prompt user for input and verify */
    printf("Enter average wind speed: ");
    scanf("%lf",&ave_wind);
    printf("Enter minimum gust: ");
    scanf("%lf",&gust_min);
    printf("Enter maximum gust (>minimum): ");
    scanf("%lf",&gust_max);

    /*  Compute wind speeds.  */
    for (timer=START_TIME; timer<=STOP_TIME; timer+=DELTA_TIME)
    {
        if (storm_flag == STORM)
            /* There is a storm, is it time to stop?  */
            if (storm_duration < MAX_DURATION)
                storm_duration += DELTA_TIME;
            else
                storm_flag = NO_STORM;
        else
            /* No storm raging, is it time for another? */
            if (rand_float(MIN_PROB,MAX_PROB) <= STORM_PROB)
            {
                storm_flag = STORM;
                storm_duration = 0;
            }
```

```
          windspeed = rand_float(ave_wind+gust_min,ave_wind+gust_max);

          if (storm_flag == STORM)
             windspeed += STORM_WIND;

          fprintf(wind_data,"%i %f\n",timer,windspeed);
       }

    /*  Close file and exit program.  */
    fclose(wind_data);
    return EXIT_SUCCESS;
}
/*---------------------------------------------------------------------*/
            (rand_float function from page 170)
/*---------------------------------------------------------------------*/

/*---------------------------------------------------------------------*/
/*    Problem chapter4_11                                              */
/*                                                                     */
/*    This program generates wind speeds with a 1% possibility of      */
/*    encountering a microburst at each time step in a small storm.    */

#include <stdio.h>
#include <stdlib.h>
#define TRUE 1
#define FALSE 0
#define DELTA_TIME 10
#define START_TIME 0
#define STOP_TIME 3600
#define FILENAME "wind.dat"
#define STORM_WIND 10
#define BURST_WIND 50
#define MIN_PROB 0
#define MAX_PROB 1
#define STORM_PROB 0.005
#define STORM_TIME 300
#define BURST_PROB 0.01
#define BURST_TIME 60

main()
{
    /*  Define variables and function prototypes.  */
    int timer=START_TIME, storm_duration=0, burst_duration=0,
        storm=FALSE, burst=FALSE;
    double ave_wind=0.0, gust_min=0.0, gust_max=0.0, windspeed=0.0;
    FILE *wind_data;
    double rand_float(double a, double b);

    /*  Open output file.  */
    wind_data = fopen(FILENAME,"w");

    /*  Prompt user for input and verify.  */
    printf("Enter average wind speed: ");
    scanf("%lf",&ave_wind);
    printf("Enter minimum gust: ");
    scanf("%lf",&gust_min);
    printf("Enter maximum gust (>minimum gust): ");
    scanf("%lf",&gust_max);

    /*  Compute wind speeds.  */
```

```
    for (timer = START_TIME; timer <= STOP_TIME; timer+=DELTA_TIME)
    {
        if (storm == TRUE)
            /* There is a storm, is it time to stop?*/
            if (storm_duration < STORM_TIME)
            {
                storm_duration += DELTA_TIME;

                /* Is there a burst? */
                if (burst == TRUE)
                    /* we have a burst in progress */
                    if (burst_duration < BURST_TIME)
                        burst_duration += DELTA_TIME;
                    else
                        burst = FALSE;
                else
                    /* no burst, but is it time for one? */
                    if (rand_float(MIN_PROB,MAX_PROB) <= BURST_PROB)
                    {
                        /* Time for a burst! */
                        burst = TRUE;
                        burst_duration = 0;
                    }
            }
            else
            {
                storm = FALSE;
                burst = FALSE;
            }
        else
            /* No storm raging, is it time for another? */
            if (rand_float(MIN_PROB,MAX_PROB) <= STORM_PROB)
            {
                storm = TRUE;
                storm_duration = 0;
                if (rand_float(MIN_PROB,MAX_PROB) <= BURST_PROB)
                {
                    burst = TRUE;
                    burst_duration = 0;
                }
            }

        windspeed = rand_float(ave_wind+gust_min,ave_wind+gust_max);

        if (storm == TRUE)
        {
            windspeed += STORM_WIND;
            if (burst == TRUE)
                windspeed += BURST_WIND;
        }

        fprintf(wind_data,"%i %f\n",timer,windspeed);
    }

    /*  Close file and exit program.  */
    fclose(wind_data);
    return EXIT_SUCCESS;
}
/*-------------------------------------------------------------------*/
            (rand_float function from page 170)
```

47

```
/*----------------------------------------------------------------*/

/*----------------------------------------------------------------*/
/*   Problem chapter4_12                                          */
/*                                                                */
/*   This program generates a file of wind data.  It allows       */
/*   the user to enter the possibility of encountering a storm.   */

#include <stdio.h>
#include <stdlib.h>
#define DELTA_TIME 10
#define START_TIME 0
#define STOP_TIME 3600
#define FILENAME "wind.dat"
#define STORM_WIND 10
#define MIN_PROB 0
#define MAX_PROB 1
#define STORM 1
#define NO_STORM 0
#define MAX_DURATION 300

main()
{
   /*  Define variables and function prototypes.   */
   int timer=START_TIME, storm_duration=0, storm_flag=NO_STORM;
   double ave_wind=0.0, gust_min=0.0, gust_max=0.0, windspeed=0.0,
          storm_prob=0;
   FILE *wind_data;
   double rand_float(double a, double b);

   /*  Open output file.   */
   wind_data = fopen(FILENAME,"w");

   /*  Prompt user for input and verify.   */
   printf("Enter average wind speed (>0): ");
   scanf("%lf",&ave_wind);
   printf("Enter minimum gust: ");
   scanf("%lf",&gust_min);
   printf("Enter maximum gust (>minimum gust): ");
   scanf("%lf",&gust_max);
   printf("Enter possiblity (percent) of a storm (0 to 100): ");
   scanf("%lf",&storm_prob);

   /*  Compute probability of a storm.   */
   storm_prob = storm_prob/100;

   /*  Compute wind speeds.   */
   for (timer=START_TIME; timer <= STOP_TIME; timer+=DELTA_TIME)
   {
      if (storm_flag == STORM)
         /* There is a storm, is it time to stop?*/
         if (storm_duration < MAX_DURATION)
            storm_duration += DELTA_TIME;
         else
            storm_flag = NO_STORM;
      else
         /* No storm raging, is it time for another? */
         if (rand_float(MIN_PROB,MAX_PROB) <= storm_prob)
         {
            storm_flag = STORM;
```

```
            storm_duration = 0;
         }

      windspeed = rand_float(ave_wind+gust_min,ave_wind+gust_max);

      if (storm_flag == STORM)
         windspeed += STORM_WIND;

      fprintf(wind_data,"%i %f\n",timer,windspeed);
   }

   /*  Close file and exit program.  */
   fclose(wind_data);
   return EXIT_SUCCESS;
}
/*-------------------------------------------------------------------*/
            (rand_float function from page 170)
/*-------------------------------------------------------------------*/

/*-------------------------------------------------------------------*/
/*  Problem chapter4_13                                              */
/*                                                                   */
/*  This program generates a wind speed data file.  The user         */
/*  enters the length in minutes for the duration of a storm.        */

#include <stdio.h>
#include <stdlib.h>
#define DELTA_TIME 10
#define START_TIME 0
#define STOP_TIME 3600
#define FILENAME "wind.dat"
#define STORM_WIND 10
#define MIN_PROB 0
#define MAX_PROB 1
#define STORM_PROB 0.005
#define STORM 1
#define NO_STORM 0


main()
{
   /*  Define variables and function prototypes.  */
   int timer=START_TIME, storm_duration=0, storm_flag=NO_STORM,
      length_storm;
   double ave_wind=0.0, gust_min=0.0, gust_max=0.0, windspeed=0.0;
   FILE *wind_data;
   double rand_float(double a, double b);

   /*  Prompt user for input and verify.  */
   printf("Enter average wind speed (>0): ");
   scanf("%lf",&ave_wind);
   printf("Enter minimum gust: ");
   scanf("%lf",&gust_min);
   printf("Enter maximum gust (> minimum gust): ");
   scanf("%lf",&gust_max);
   printf("Enter storm length in minutes: ");
   scanf("%lf",&length_storm);
   length_storm = length_storm * 60;

   /*  Open output file.  */
```

```
   wind_data = fopen(FILENAME,"w");

   /*  Compute wind speeds.  */
   for (timer = START_TIME; timer <= STOP_TIME; timer+=DELTA_TIME)
   {
      if (storm_flag == STORM)
         /* There is a storm, is it time to stop?*/
         if (storm_duration < length_storm)
            storm_duration += DELTA_TIME;
         else
            storm_flag = NO_STORM;
      else
         /* No storm raging, is it time for another? */
         if (rand_float(MIN_PROB,MAX_PROB) <= STORM_PROB)
         {
            storm_flag = STORM;
            storm_duration = 0;
         }

      windspeed = rand_float(ave_wind+gust_min,ave_wind+gust_max);

      if (storm_flag == STORM)
         windspeed += STORM_WIND;

      fprintf(wind_data,"%i %f\n",timer,windspeed);
   }

   /*  Close the file and exit program. */
   fclose(wind_data);
   return EXIT_SUCCESS;
}
/*------------------------------------------------------------------*/
            (rand_float function from page 170)
/*------------------------------------------------------------------*/

/*------------------------------------------------------------------*/
/*  Problem chapter4_14                                             */
/*                                                                  */
/*  This program generates a wind speed file with storms whose      */
/*  length is a random number that varies between 3 and 5 minutes.  */

#include <stdio.h>
#include <stdlib.h>
#define DELTA_TIME 10
#define START_TIME 0
#define STOP_TIME 3600
#define FILENAME "wind.dat"
#define STORM_WIND 10
#define MIN_PROB 0
#define MAX_PROB 1
#define STORM_PROB 0.005
#define STORM 1
#define NO_STORM 0
#define MIN_LENGTH 3
#define MAX_LENGTH 5

main()
{
   /*  Define variables and function prototypes.  */
   int timer=START_TIME, storm_duration=0, storm_flag=NO_STORM;
```

50

```
   double ave_wind=0.0, gust_min=0.0, gust_max=0.0, windspeed=0.0,
          storm_length=0;
   FILE *wind_data;
   double rand_float(double a, double b);

   /*  Open output file.  */
   wind_data = fopen(FILENAME,"w");

   /*  Prompt user for input and verify.  */
   printf("Enter average wind speed: (>0)");
   scanf("%lf",&ave_wind);
   printf("Enter minimum gust: ");
   scanf("%lf",&gust_min);
   printf("Enter maximum gust: (> minimum gust)");
   scanf("%lf",&gust_max);

   /*  Compute wind speeds.  */
   for (timer=START_TIME; timer <= STOP_TIME; timer+=DELTA_TIME)
   {
      if (storm_flag == STORM)
         /* There is a storm, is it time to stop?*/
         if (storm_duration < storm_length)
            storm_duration += DELTA_TIME;
         else
            storm_flag=NO_STORM;
      else
         /* No storm raging, is it time for another? */
         if (rand_float(MIN_PROB,MAX_PROB) <= STORM_PROB)
         {
            storm_flag = STORM;
            storm_duration = 0;
            /* get the length of this storm in seconds */
            storm_length = 60.0 * rand_float(MIN_LENGTH,MAX_LENGTH);
         }

      windspeed = rand_float(ave_wind+gust_min,ave_wind+gust_max);

      if (storm_flag == STORM)
         windspeed += STORM_WIND;

      fprintf(wind_data,"%i %f\n",timer,windspeed);
   }

   /*  Close file and exit program.  */
   fclose(wind_data);
   return EXIT_SUCCESS;
}
/*--------------------------------------------------------------------*/
            (rand_float function from page 170)
/*--------------------------------------------------------------------*/


/*--------------------------------------------------------------------*/
/*  Problem chapter4_15                                               */
/*                                                                    */
/*  This program determines the real roots of a quadratic            */
/*  equation, assuming that the user enters the coefficients of      */
/*  the quadratic equation.                                          */

#include <stdio.h>
#include <stdlib.h>
```

```
#include <math.h>

main()
{
   /*   Declare variables.   */
   double a, b, c, discriminant, root1,root2;

   /*   Prompt user for equation.   */
   printf("Enter a,b,c for equation ax^2 + bx + c: ");
   scanf("%lf %lf %lf",&a,&b,&c);
   printf("Equation is: %fx^2 + %fx + %f = y\n",a,b,c);

   /*   Are the roots complex?   */
   discriminant = b*b - 4*a*c;
   if (discriminant < 0.0)
      printf("Complex roots!");
   else
   {
      root1 = (-b + sqrt(discriminant)) / (2*a);
      root2 = (-b - sqrt(discriminant)) / (2*a);
      printf("Roots are: x1=%f, x2=%f\n",root1,root2);
    }

   /*   Exit program.   */
   return EXIT_SUCCESS;
}
/*-------------------------------------------------------------------------*/

/*-------------------------------------------------------------------------*/
/*   Problem chapter4_16                                                   */
/*                                                                         */
/*   This program determines the roots of a quadratic equation.           */

#include <stdio.h>
#include <stdlib.h>
#include <math.h>

main()
{
   /*   Declare variables.   */
   double a, b, c, discriminant, root1, root2;

   /*   Prompt user for equation.   */
   printf("Enter a,b,c for equation ax^2 + bx + c: ");
   scanf("%lf %lf %lf",&a,&b,&c);
   printf("Equation is: %fx^2 + %fx + %f = y\n",a,b,c);

   /*   Are the roots complex?   */
   discriminant = b*b - 4*a*c;
   if (discriminant < 0.0)
   {
      root1 = -b/(2*a);
      root2 = sqrt(abs(discriminant))/(2*a);
      printf("Roots are: x1= %f + j%f, x2 = %f -j%f\n",
             root1 root2,root1,root2);
    }
   else
   {
      root1 = (-b + sqrt(discriminant)) / (2*a);
      root2 = (-b - sqrt(discriminant)) / (2*a);
```

```
            printf("Roots are: x1=%f, x2=%f\n",root1,root2);
    }

    /*  Exit program.  */
    return EXIT_SUCCESS;
}
/*------------------------------------------------------------------*/

/*------------------------------------------------------------------*/
/*  Problem chapter4_17                                             */
/*                                                                  */
/*  This program evaluates this mathematical function:              */
/*       f(x) = 0.1 x^2 - x ln x                                    */

#include <stdio.h>
#include <stdlib.h>
#include <math.h>

main()
{
    /*  Declare variables and function prototypes.  */
    int n, k;
    double a, b, step, left, right;
    double f(double x);
    void check_roots(double left, double right);

    /*  Get user input.  */
    printf("Enter interval limits a, b (a<b): \n");
    scanf("%lf %lf",&a,&b);
    printf("Enter step size: \n");
    scanf("%lf",&step);

    /*  Check subintervals for roots.  */
    n = ceil((b - a)/step);
    printf("n is %i\n",n);
    for (k=0; k<=n-1; k++)
    {
        left = a + k*step;
        if (k == n-1)
            right = b;
        else
            right = left + step;
        check_roots(left,right);
    }
    check_roots(b,b);

    /*  Exit program.  */
    return EXIT_SUCCESS;
}
/*------------------------------------------------------------------*/
/*  This function checks a subinterval for a root.                  */

void check_roots(double left, double right)
{
    /*  Declare variables and function prototypes.  */
    double f_left, f_right;
    double f(double x);

    /*  Evaluate subinterval endpoints and  test for roots.  */
    f_left = f(left);
```

53

```
      f_right = f(right);
      if (fabs(f_left) < 0.1e-04)
         printf("Root detected at %.3f \n",left);
      else
         if (fabs(f_right) < 0.1e-04)
            ;
         else
            if (f_left*f_right < 0)
               printf("Root detected at %.3f \n",(left+right)/2);

      /*  Exit function.  */
      return;
}
/*-------------------------------------------------------------------*/
/*  This functions evaluates a mathematical function given           */
/*  in problem 17. Be sure not to call with x < 0.                   */

double f(double x)
{
   /*  Return function value.  */
   return (0.1*x*x - x*log(x));
}
/*-------------------------------------------------------------------*/

/*-------------------------------------------------------------------*/
/*  Problem chapter4_18                                              */
/*                                                                   */
/*  This program finds the roots of this function in a               */
/*  user-specified interval:  f(x) = sinc(x)                         */

#include <stdlib.h>
#include <math.h>

main()
{
   /*  Declare variables and function prototypes.   */
   int n, k;
   double a, b, step, left, right;
   void check_roots(double left, double right);

   /*  Get user input.  */
   printf("Enter interval limits a, b (a<b): \n");
   scanf("%lf %lf",&a,&b);
   printf("Enter step size: \n");
   scanf("%lf",&step);

   /*  Check subintervals for roots.  */
   n = ceil((b - a)/step);
   for (k=0; k<=n-1; k++)
   {
      left = a + k*step;
      if (k == n-1)
         right = b;
      else
         right = left + step;
      check_roots(left,right);
   }
   check_roots(b,b);

   /*  Exit program.  */
```

```c
      return EXIT_SUCCESS;
}
/*--------------------------------------------------------------------*/
/*  This function checks a subinterval for a root.                   */

void check_roots(double left, double right)
{
   /*  Declare variables and function prototypes.   */
   double f_left, f_right;
   double sinc(double x);

   /*  Evaluate subinterval endpoints and  */
   /*  test for roots.                      */
   f_left = sinc(left);
   f_right = sinc(right);
   if (fabs(f_left) < 0.1e-04)
      printf("Root detected at %.3f \n",left);
   else
      if (fabs(f_right) < 0.1e-04)
         ;
      else
         if (f_left*f_right < 0)
            printf("Root detected at %.3f \n",
               (left+right)/2);

   /*  Void return.  */
   return;
}
/*--------------------------------------------------------------------*/
/* This function evaluates a sinc function.                          */

double sinc(double x)
{
   /*  Return function value.  */
   if (x == 0)
      return 1;
   else
      return sin(x)/x;
}
/*--------------------------------------------------------------------*/

/*--------------------------------------------------------------------*/
/*  Problem chapter4_19                                              */
/*                                                                   */
/*  This program estimates the roots of a subinterval using this     */
/*  approximation:  c = (a*f(b) - b*f(b)) / (f(b) - f(a))            */

#include <stdio.h>
#include <stdlib.h>
#include <math.h>

main()
{
   /*  Declare variables and function prototypes.  */
   int n, k;
   double a0, a1, a2, a3, a, b, step, left, right;
   void check_roots(double left, double right, double a0,
                 double a1, double a2, double a3);

   /*  Get user input.  */
```

55

```
   printf("Enter coefficients a0, a1, a2, a3: \n");
   scanf("%lf %lf %lf %lf",&a0,&a1,&a2,&a3);
   printf("Enter interval limits a, b (a<b): \n");
   scanf("%lf %lf",&a,&b);
   printf("Enter step size: \n");
   scanf("%lf",&step);

   /*  Check subintervals for roots.  */
   n = ceil((b - a)/step);
   for (k=0; k<=n-1; k++)
   {
      left = a + k*step;
      if (k == n-1)
         right = b;
      else
         right = left + step;
      check_roots(left,right,a0,a1,a2,a3);
   }
   check_roots(b,b,a0,a1,a2,a3);

   /*  Exit program.  */
   return EXIT_SUCCESS;
}
/*-------------------------------------------------------------*/
/*  This function checks a subinterval for a root.            */

void check_roots(double left, double right, double a0,
                 double a1, double a2, double a3)
{
   /*  Declare variables and function prototypes.  */
   double f_left, f_right,c;
   double polynom(double x, double a0, double a1,
                  double a2, double a3);

   /*  Evaluate subinterval endpoints and test for roots.  */
   f_left = polynom(left,a0,a1,a2,a3);
   f_right = polynom(right,a0,a1,a2,a3);

   if (fabs(f_left) < 0.1e-04)
      printf("Root detected at %.3f \n",left);
   else
      if (fabs(f_right) < 0.1e-04)
         ;
      else
         if (f_left*f_right < 0)
         {
            c = (left*f_right - right*f_left)/(f_right - f_left);
            printf("Root detected at %.3f \n",c);
         }

   /*  Void return.  */
   return;
}
/*-------------------------------------------------------------*/
            (poly function from page 187)
/*-------------------------------------------------------------*/

/*-------------------------------------------------------------*/
/*  Problem chapter4_20                                        */
/*
```

```c
/*  This program computes an approximation to a factorial using    */
/*   Stirling formula:  n! = sqrt(2*pi*n) (/3)^n                    */

#include <stdio.h>
#include <stdlib.h>
#include <math.h>

#define EULER 2.718282
#define PI    3.141593

main()
{
    /*  Define variables and function prototypes.  */
    int n;
    int n_fact(int n);

    /*  Get n from user.  */
    printf("Enter n: ");
    scanf("%i",&n);
    while (n < 0)
    {
       printf("n must not be negative\n");
       printf("Enter n: ");
       scanf("%i",&n);
    }

    /*  Calculate the Stirling approximation for integers.  */
    printf(" n! is about %i \n",n_fact(n));

    /*  Exit program.  */
    return EXIT_SUCCESS;
}
/*--------------------------------------------------------------------*/
/*  This function calculates a Stirling approximation for integers.   */
/*  Answers are rounded to nearest integer, not truncated.            */

int n_fact(int n)
{
    /*  Return approximation value.  */
    if (n == 0)
       return 0;
    else
       return ((int)(sqrt(2*PI*n)*pow((n/EULER),n) + 0.5));
}
/*--------------------------------------------------------------------*/

/*--------------------------------------------------------------------*/
/*  Function chapter4_21                                              */
/*                                                                   */
/*  This function calculates the number of permutations of n things, */
/*   taken k at a time.                                              */

int permute(int n, int k)
{
    /*  Declare variables.  */
    int n_fact(int);

    /*  Return number of permuations.  */
    return (int)(n_fact(n)/(n_fact(n-k)) + 0.5 );
}
```

```
/*------------------------------------------------------------------*/
               (n_fact function from Problem 4_20)
/*------------------------------------------------------------------*/


/*------------------------------------------------------------------*/
/*  Function chapter4_22                                            */
/*                                                                  */
/*  This function calculates the number of combinations of n things, */
/*  taken k at a time using a factorial function from problem 20.    */

int combine(int n, int k)
{
   /*  Define function prototype.  */
   int n_fact(int);

   /*  Return number of combinations.  */
   return (int)(n_fact(n)/(n_fact(k)*n_fact(n-k)) + 0.5 );
}
/*------------------------------------------------------------------*/
               (n_fact function from Problem 4_20)
/*------------------------------------------------------------------*/


/*------------------------------------------------------------------*/
/*  Problem chapter4_23                                             */
/*                                                                  */

/*  This program compares the cosine of an angle using the library  */
/*  function with the value computed from the first five terms      */
/*  of this series:  cos x = 1 - x^2/2! + x^4/4! .....              */

#include <stdio.h>
#include <stdlib.h>
#include <math.h>

#define EULER 2.718282
#define PI 3.141593
#define SERIES_LENGTH 5

main()
{
   /*  Define variables and function prototypes.  */
   int n_fact(int n);
   double x;
   double s_cosine(double x);

   /*  Get x from user.  */
   printf("\nEnter x in radians: ");
   scanf("%lf",&x);

   /*  Get series cosine.  */
   printf("Series cosine: %f\n",s_cosine(x));
   printf("C library cosine: %f\n",cos(x));

   /*  Exit program.  */
   return EXIT_SUCCESS;
}
/*------------------------------------------------------------------*/
/* This function calculates the series for cosine                   */
/* for a specified number of terms.                                 */
```

```
double s_cosine(double x)
{
    /*  Declare variables and function prototypes.  */
    int i;
    double sum=1;
    int n_fact(int);

    /*  Compute cosine sum.  */
    for (i=1; i<SERIES_LENGTH; i++)
    {
        sum += pow(-1,i)*pow(x,2*i)/n_fact(2*i);
    }

    /*  Return series sum.  */
    return sum;
}
/*----------------------------------------------------------------*/
            (n_fact function from Problem 4_20)
/*----------------------------------------------------------------*/

/*----------------------------------------------------------------*/
/*  Problem chapter4_24                                           */
/*                                                                */
/*  This program compares the cosine of an angle using the library */
/*  function with the value computed from terms > 0.0001          */
/*  of this series:  cos x = 1 - x^2/2! + x^4/4! .....            */

#include <stdio.h>
#include <stdlib.h>
#include <math.h>

#define EULER 2.718282
#define PI 3.141593
#define MAX_TERM 0.001

main()
{
    /*  Define variables and function prototypes.  */
    unsigned int n_fact(int n);
    double x;
    double s_cosine(double);

    /*  Get x from user.  */
    printf("\nEnter x in radians: ");
    scanf("%lf",&x);

    /*  Get series cosine.  */
    printf("Series cosine: %f\n",s_cosine(x));
    printf("C library cosine: %f\n",cos(x));

    /*  Exit program. */
    return EXIT_SUCCESS;
}
/*----------------------------------------------------------------*/
/*  This function caluclates the series for cosine using terms    */
/*  that are greater than a specified value.                      */

double s_cosine(double x)
{
    /*  Declare variables and function prototypes.  */
```

```c
   int i=0;
   double sum=0, term=1;
   unsigned int n_fact(int);

   /*  Determine cosine sum.  */
   while (fabs(term) > MAX_TERM)
   {
      sum  += term;
      i++;
      term = pow(-1,i)*pow(x,2*i)/n_fact(2*i);
   }
   printf("Number of terms in the series: %i\n",i);

   /*  Return cosine sum.  */
   return sum;
}
/*-------------------------------------------------------------------*/
            (n_fact function from Problem 4_20)
/*-------------------------------------------------------------------*/
```

Chapter 5

```
/*------------------------------------------------------------------*/
/*   Problem chapter5_1                                             */
/*                                                                  */
/*   This program reads the wind-tunnel test data, and then allows  */
/*   the user to enter a flight-path angle.  If the angle is within */
/*   the bounds of the data set, the program then uses linear       */
/*   interpolation to compute the corresponding coefficient of lift. */

#include <stdio.h>
#include <stdlib.h>
#define FILENAME "tunnel.dat"
#define MAX_INDEX 20

main()
{
   /*  Define variables and function prototypes.  */
   int index=0, angle[MAX_INDEX];
   double coef[MAX_INDEX], request;
   FILE *tunnel;
   double interpolate(double, int x[], double y[]);

   /*  Open input file.  */
   tunnel = fopen(FILENAME,"r");

   /*  Read data.  */
   while ((fscanf(tunnel,"%i %lf",&angle[index],&coef[index])) == 2)
      index++;

   /*  Prompt user for input.  */
   printf("Enter angle: ");
   scanf("%lf",&request);

   /*  Interpolate the coefficent of lift, if angle in range.  */
   if ( (request < angle[0]) || (request > angle[index-1]))
      printf("Angle out of range ,cannot interpolate.\n");
   else
      printf("Coefficient of lift (interpolated): %5.3f\n",
             interpolate(request ,angle ,coef));

   /*  Exit program.  */
   return EXIT_SUCCESS;
}
/*------------------------------------------------------------------*/
/*   This function interpolates y values, given x and a new value.  */
/*   It is assumed that x is in ascending order.                    */

double interpolate(double value, int x[], double y[])
{
   /*  Declare variables.  */
   int i=0;

   /*  Find the angles that value is between.  */
   while (value > (double)x[i])
      i++;

   /*  Return interpolated value.  */
   if (value == (double) x[i])
```

```
            return y[i];
        else
            /*  value is between x[i-1] and x[i]  */
            return (y[i-1] + (value - x[i-1])/
                    (x[i]-x[i-1]) * (y[i] - y[i-1]));
}
/*-------------------------------------------------------------------*/


/*-------------------------------------------------------------------*/
/*  Problem chapter5_2                                              */
/*                                                                  */
/*  This program reads the wind-tunnel test data, and prints the    */
/*  range of angles covered in the file.  It then allows            */
/*  the user to enter a flight-path angle.  If the angle is within  */
/*  the bounds of the data set, the program then uses linear        */
/*  interpolation to compute the corresponding coefficient of lift. */

#include <stdio.h>
#include <stdlib.h>
#define FILENAME "tunnel.dat"
#define MAX_INDEX 20

main()
{
    /*  Define variables and function prototypes.  */
    int index=0, angle[MAX_INDEX];
    double coef[MAX_INDEX], request;
    FILE *tunnel;
    double interpolate(double, int x[], double y[]);

    /*  Open input file.  */
    tunnel = fopen(FILENAME,"r");

    /*  Read data.  */
    while ((fscanf(tunnel,"%i %lf",&angle[index],&coef[index])) == 2)
        index++;

    /*  Tell user the range of angles.  */
    printf("Angles range from %i to %i\n",angle[0],angle[index-1]);

    /*  Prompt user for input.  */
    printf("Enter angle: ");
    scanf("%lf",&request);


    /*  Interpolate the coefficent of lift, if angle in range.  */
    if ((request<angle[0]) || (request>angle[index-1]))
        printf("Angle out of range , cannot interpolate.\n");
    else
        printf("Coefficient of lift (interpolated): %5.3f\n",
                interpol(request ,angle, coef));

    /*  Exit program.  */
    return EXIT_SUCCESS;
}
/*-------------------------------------------------------------------*/
            (interpolate function from Problem 5-1)
/*-------------------------------------------------------------------*/


/*-------------------------------------------------------------------*/
```

```
/*   Function chapter5_3                                              */
/*                                                                    */
/*   This function verifies that the flight-path angles are in        */
/*   ascending order.  The function returns a zero if the angles      */
/*   are not in order and a 1 if they are in order.                   */

int ordered(double x[], int num_pts)
{

   #define ASCENDING 1
   #define NOT_ASCENDING 0

   /*   Declare variables.   */
   int i;

   /*   Determine if angles are out of order.   */
   for (i=1; i<num_pts; i++)
      if ( x[i-1] > x[i] )
            return NOT_ASCENDING;

   /*   Return in order value.   */
   return ASCENDING;
}
/*------------------------------------------------------------------*/


/*------------------------------------------------------------------*/
/*   Function chapter5_4                                              */
/*                                                                    */
/*   This function reorders the values in x so that they are          */
/*   in ascending order and the relationship between x and y is       */
/*   preserved.                                                       */

void reorder(double x[], double y[], int num_pts)
{
   /*   Declare variables.   */
   int m, j, k;
   double hold;

   /*   Reorder array values.   */
   for (k=0; k<num_pts-1; k++)
   {
      /*   Exchange minimum with next array value. */
      m = k;
      for (j=k+1; j<num_pts; j++)
      {
         if (x[j] < x[m])
            m=j;
      }
      hold = x[m];
      x[m] = x[k];
      x[k] = hold;
      hold = y[m];
      y[m] = y[k];
      y[k] = hold;
   }

   /*   void return */
   return;
}
/*------------------------------------------------------------------
```

63

```
/*---------------------------------------------------------------*/
/*   Problem chapter5_5                                          */
/*                                                              */
/*   This program uses the function developed in problem 3
/*   to determine whether or not the data are in the desired order.
/*   If they are not in the desired order, it uses the function
/*   developed in problem 4 to reorder them. It then performs a
/*   linear interpolation to determine new coefficients of lift.  */

#include <stdio.h>
#include <stdlib.h>
#define FILENAME "tunnel.dat"
#define MAX_INDEX 20

main()
{
    /*   Define variables and function prototypes.   */
    int index=0;
    double angle[MAX_INDEX], coef[MAX_INDEX], request;
    FILE *tunnel;
    double interpolate(double, double x[],double y[]);
    int ordered(double x[], int num_pts);
    void reorder(double x[], double y[], int num_pts);

    /*   Open input file.   */
    tunnel = fopen(FILENAME,"r");

    /*   Read data.   */
    while ((fscanf(tunnel,"%lf %lf",&angle[index],&coef[index])) == 2)
        index++;

    /*   Reorder data if necessary.   */
    if (ordered(angle,index) == ASCENDING)
        printf("Flight-path angles are in ascending order.\n");
    else
    {
        printf("Out of order flight-path angles are being reordered.\n");
        reorder(angle,coef,index);
    }

    /*   Tell user the range of angles.   */
    printf("Angles range from %3.2f to %3.2f\n",
            angle[0],angle[index-1]);

    /*   Prompt user for input.   */
    printf("Enter angle: ");
    scanf("%lf",&request);

    /*   Interpolate the coefficent of lift, if angle in range.   */
    if ( (request < angle[0]) || (request > angle[index-1]))
        printf("Angle out of range, cannot interpolate.\n");
    else
        printf("Coefficient of lift (interpolated): %5.3f\n",
                interpolate(request,angle,coef));

    /*   Exit program.   */
    return EXIT_SUCCESS;
/*---------------------------------------------------------------*/
            (ordered function from Problem 5-3)
```

64

```
/*------------------------------------------------------------------*/
            (reorder function from Problem 5-4)
/*------------------------------------------------------------------*/
            (interpolate function from Problem 5-1)
/*------------------------------------------------------------------*/


/*------------------------------------------------------------------*/
/*   Problem chapter5_6                                             */
/*                                                                  */
/*   This program generates sequences of random floating-point      */
/*   values between 4 and 10.  It then compare the computed mean     */
/*   and variance to the theoretical values.                        */

#include <stdio.h>
#include <stdlib.h>
#include <math.h>
#define MIN_NUM 4.0
#define MAX_NUM 10.0
#define MAX_TIMES 5000

main()
{
    /*   Define variables and function prototypes. */
    int times=0, index=0;
    double number[MAX_TIMES], sample_mean, var_sample;
    double rand_float(double, double);
    double variance(double x[], int);
    double mean(double x[], int);

    /*   Get number of samples from user. */
    printf("Enter number of random numbers to use (< %i): ",MAX_TIMES);
    scanf("%i",&times);

    /*   Generate random numbers in the sample. */
    for (index=0; index<times; index++)
       number[index] = rand_float(MIN_NUM,MAX_NUM);

    /*   Calculate sample mean and variance */
    var_sample = variance(number,times);
    sample_mean = mean(number,times);

    /*   Print results. */
    printf("Sample mean: %f.  Sample variance: %f.\n",
           sample_mean,var_sample);

    /*   Exit program.   */
    return EXIT_SUCCESS;
}
/*------------------------------------------------------------------*/
            (rand_float function from page 170)
/*------------------------------------------------------------------*/
            (mean function from page 217)
/*------------------------------------------------------------------*/
            (variance function from page 220)
/*------------------------------------------------------------------*/


/*------------------------------------------------------------------*/
/*   Problem chapter5_7                                             */
/*                                                                  */
/*                                                                  */
```

```c
/*   This program generates two sequences of 500 points.   Each          */
/*   sequence should have a theoretical mean of 4, but one sequence      */
/*   should have a variance of 0.5 and the other a variance of 2.        */
/*   The computed means and variances are printed.                       */

#include <stdio.h>
#include <stdlib.h>
#include <math.h>
#define MEAN 4.0
#define VAR1 0.5
#define VAR2 2.0
#define MAX_POINTS 500

main()
{
    /*   Define variables and function prototypes.   */
    int index = 0;
    double first_sequence[MAX_POINTS], second_sequence[MAX_POINTS];
    double a1, b1, a2, b2, mean1, mean2, var1, var2;
    double rand_float(double, double);
    double variance(double x[], int);
    double mean(double x[], int);

    /*   Find the limits on the random number generator.   */
    a1 = (2*MEAN - sqrt(12*VAR1))/2;
    b1 = 2*MEAN - a1;
    a2 = (2*MEAN - sqrt(12*VAR2))/2;
    b2 = 2*MEAN - a2;

    /*   Generate random numbers in the sample.   */
    for (index=0; index<MAX_POINTS; index++)
    {
        first_sequence[index] = rand_float(a1,b1);
        second_sequence[index] = rand_float(a2,b2);
    }

    /*   Calculate sample means and variances.   */
    var1 = variance(first_sequence,MAX_POINTS);
    var2 = variance(second_sequence,MAX_POINTS);
    mean1 = mean(first_sequence,MAX_POINTS);
    mean2 = mean(second_sequence,MAX_POINTS);

    /*   Print results.   */
    printf("Sequence 1:  Mean: %f  Variance: %f.\n",mean1,var1);
    printf("Sequence 2:  Mean: %f  Variance: %f.\n",mean2,var2);
    printf("Theoretically: Mean: %f Variance 1: %f Variance 2: %f.\n",
           MEAN,VAR1,VAR2);

    /*   Exit program.   */
    return EXIT_SUCCESS;
}
/*-----------------------------------------------------------------------*/
            (rand_float function from page 170)
/*-----------------------------------------------------------------------*/
            (mean function from page 217)
/*-----------------------------------------------------------------------*/
            (variance function from page 220)
/*-----------------------------------------------------------------------*/

/*-----------------------------------------------------------------------*/
```

```
/*   chapter5_8                                                          */
/*                                                                       */
/*   This program generates two sequences of 500 points.  Each           */
/*   sequence should have a theoretical variance of 3, but one           */
/*   sequence should have a mean of 0 and the other a mean of -4.        */
/*   The computed means and variances are printed.                       */

#include <stdio.h>
#include <stdlib.h>
#include <math.h>
#define MEAN1 0.0
#define MEAN2 -4.0
#define VAR 3.0
#define MAX_POINTS 500

main()
{
   /*   Define variables and function prototypes.  */
   int index=0;
   double first_sequence[MAX_POINTS], second_sequence[MAX_POINTS];
   double a1, b1, a2, b2, mean1, mean2, var1, var2;
   double rand_float(double, double);
   double variance(double x[], int);
   double mean(double x[], int);

   /*   Find the limits on the random number generator.  */
   a1 = (2*MEAN1 - sqrt(12*VAR))/2;
   b1 = 2*MEAN1 - a1;
   a2 = (2*MEAN2 - sqrt(12*VAR))/2;
   b2 = 2*MEAN2 - a2;

   /*   Generate random numbers in the sample.  */
   for (index=0; index<MAX_POINTS; index++)
   {
      first_sequence[index] = rand_float(a1,b1);
      second_sequence[index] = rand_float(a2,b2);
   }

   /*   Calculate sample means and variances.  */
   var1 = variance(first_sequence,MAX_POINTS);
   var2 = variance(second_sequence,MAX_POINTS);
   mean1 = mean(first_sequence,MAX_POINTS);
   mean2 = mean(second_sequence,MAX_POINTS);

   /*   Print results.  */
   printf("Sequence 1:  Mean: %f  Variance: %f.\n",mean1,var1);
   printf("Sequence 2:  Mean: %f  Variance: %f.\n",mean2,var2);
   printf("Theoretically: Mean 1: %f Mean 2: %f Variance: %f.\n",
          MEAN1,MEAN2,VAR);

   /*   Exit program.  */
   return EXIT_SUCCESS;
}

/*----------------------------------------------------------------------*/
            (rand_float function from page 170)
/*----------------------------------------------------------------------*/
            (mean function from page 217)
/*----------------------------------------------------------------------*/
            (variance function from page 220)
```

```
/*------------------------------------------------------------------*/

/*------------------------------------------------------------------*/
/*  Function chapter5_9                                          */
/*                                                               */
/*  This function generates a random number                     */
/*  with a specified mean and variance.                         */

double rand_mv(double mu, double sigma_sq)
{
   /*  Declare variables.   */
   double a, b;

   /*  Find the limits on the random number generator  */
   a = (2*mu - sqrt(12*sigmasq))/2;
   b = 2*mu - a;

   /*  Return the random number.   */
   return rand_float(a,b);
}
/*------------------------------------------------------------------*/
            (rand_float function from text page xx goes here)
/*------------------------------------------------------------------*/
            (mean function from text page xx goes here)
/*------------------------------------------------------------------*/
            (variance function from text page xx goes here)
/*------------------------------------------------------------------*/

/*------------------------------------------------------------------*/
/*  Problem chapter5_10                                          */
/*                                                               */
/*  This program computes and prints the average power output over */
/*  a period of time.  It also print the number of days with    */
/*  greater-than-average power output.                          */

#include <stdio.h>
#include <stdlib.h>
#define NCOL 7
#define NROW 10
#define FILENAME "power1.dat"

main()
{
   /*  Define variables.   */
   double total_power=0, average_power, power_out[NROW][NCOL];
   unsigned int greater=0, row, col;
   FILE *power;

   /*  Open input file.   */
   power = fopen(FILENAME,"r");

   /*  Read data.   */
   for (row=0; row<NROW; row++)
      for (col=0; col<NCOL; col++)
      {
         fscanf(power,"%lf",&power_out[row][col]);
         total_power += power_out[row][col];
      }

   /*  Calculate the average power.   */
```

```c
    average_power = total_power/(NROW*NCOL);

    /*  Count the days with greater than average power output.  */
    for (row=0; row < NROW; row++)
       for (col=0; col < NCOL; col++)
          if (power_out[row][col] > average_power)
             greater++;

    /*  Print the results.  */
    printf("Average power output: %.3f \n",average_power);
    printf("Number of days with greater than average power output:"
          " %u.\n",greater);

    /*  Return */
    return EXIT_SUCCESS;
}
/*-------------------------------------------------------------------*/

/*-------------------------------------------------------------------*/
/*  Problem chapter5_11                                              */
/*                                                                   */
/*  This program reads a set of power plant output data from a       */
/*  data file.  It then prints the day of the week and the          */
/*  number of the week on which the minimum power output occurred.   */
/*  If there are several days with the minimum power output,         */
/*  it prints the information for each day.                          */

#include <stdio.h>
#include <stdlib.h>
#define NCOL 7
#define NROW 10
#define MAX_POWER 99999
#define FILENAME "power1.dat"

main()
{
    /*  Define variables.  */
    double min_power=MAX_POWER, power_out[NROW][NCOL];
    unsigned int row, col;
    FILE *power;

    /*  Open input file.  */
    power = fopen(FILENAME,"r");

    /*  Read data.  */
    for (row=0; row<NROW; row++)
       for (col=0; col<NCOL; col++)
       {
          fscanf(power,"%lf",&power_out[row][col]);
          if (power_out[row][col] < min_power)
             min_power = power_out[row][col];
       }

    /*  Find the days with minimum power output.  */
    printf("Minimum power output was: %.3f.\n",min_power);
    for (row=0; row<NROW; row++)
       for (col=0; col<NCOL; col++)
          if (power_out[row][col] == min_power)
             printf("Week %u, day %u was a minimum power out day.\n",
                   row,col);
```

```
    /*  Exit program.  */
    return EXIT_SUCCESS;
}
/*------------------------------------------------------------------*/

/*------------------------------------------------------------------*/
/*  Function chapter5_12                                            */
/*                                                                  */
/*  This function computes the average of a specified column of a   */
/*  two-dimensional array with NROWS rows and NCOLS columns.        */

double col_ave(double x[NROWS][NCOLS], int col)
{
    /*  Declare and initialize variables.  */
    int row;
    double sum=0;

    /*  Compute column sum.  */
    for (row=0; row<NROWS; row++)
       sum += x[row][col];

    /*  Return column average.  */
    return sum/NROWS;
}
/*------------------------------------------------------------------*/

/*------------------------------------------------------------------*/
/*  Problem chapter5_13                                             */
/*                                                                  */
/*  This program prints a report that lists the average power       */
/*  output for the first day of the week, and so on.                */

#include <stdio.h>
#include <stdlib.h>
#define NCOLS 7
#define NROWS 10
#define FILENAME "power1.dat"

main()
{
    /*  Define variables and function prototypes.  */
    double power_out[NROWS][NCOLS];
    int row, col;
    FILE *power;
    double col_ave(double x[][], int col);

    /*  Open input file.  */
    power = fopen(FILENAME,"r");

    /*  Read data.  */
    for (row=0; row<NROWS; row++)
       for (col=0; col<NCOLS; col++)
          fscanf(power,"%lf",&power_out[row][col]);

    /*  Find column average for a given column.  */
    for (col = 0; col<NCOLS; col++)
    {
       printf("Day %i: Average Power Output in Megawatts:  %6.2f\n",
              col,col_ave(power_out,col));
```

70

```
    }

    /*  Exit program.  */
    return EXIT_SUCCESS;
}
/*----------------------------------------------------------------*/
                (col_ave function from Problem 5-12)
/*----------------------------------------------------------------*/

/*----------------------------------------------------------------*/
/*  Function chapter 5_14                                        */
/*                                                               */
/*  This function computes the average of a specified row of a   */
/*  two-dimensional array with NROWS rows and NCOLS columns.     */

double row_ave(double x[NROWS][NCOLS], int row)
{
    /*  Declare variables.  */
    int col;
    double sum=0;

    /*  Compute row sum.  */
    for (col=0; col < NCOLS; col++)
        sum += x[row][col];

    /*  Compute row average.  */
    return sum/NCOLS;
}
/*----------------------------------------------------------------*/

/*----------------------------------------------------------------*/
/*  Problem chapter5_15                                          */
/*                                                               */
/*  This program prints a report that lists the average power    */
/*  output for the first week, the second week and so on.        */

#include <stdio.h>
#include <stdlib.h>
#define NCOLS 7
#define NROWS 10
#define FILENAME "power1.dat"

main()
{
    /*  Define variables.  */
    double power_out[NROWS][NCOLS];
    int row, col;
    FILE *power;
    double row_ave(double x[][], int row);

    /*  Open input file.  */
    power = fopen(FILENAME,"r");

    /*  Read data.  */
    for (row=0; row<NROWS; row++)
        for (col=0; col<NCOLS; col++)
            fscanf(power,"%lf",&power_out[row][col]);

    /*  Find row average for a given row. */
    for (row = 0; row<NROWS; row++)
```

71

```c
    {
       printf("Week %i: Average Power Output in Megawatts:   %6.2f\n",
              row,row_ave(power_out,row));
    }

    /*  Exit program.  */
    return EXIT_SUCCESS;
}
/*-------------------------------------------------------------------*/
                  (col_ave function from Problem 5-14)
/*-------------------------------------------------------------------*/

/*-------------------------------------------------------------------*/
/*  Problem chapter 5_16                                             */
/*                                                                   */
/*  Write a program to compute and print the mean and variance      */
/*  of the power plant output data.                                  */

#include <stdio.h>
#include <stdlib.h>
#define NCOLS 7
#define NROWS 10
#define FILENAME "power1.dat"

main()
{
    /*  Declare and initialize variables.  */
    int power_out[NROWS][NCOLS], row, col;
    double mean=0, var=0;
    FILE *power;
    double mean=0, var=0;

    /*  Open input file.  */
    power = fopen(FILENAME,"r");

    /*  Read data.  */
    for (row=0; row<NROWS; row++)
       for (col=0; col<NCOLS; col++)
       {
           fscanf(power,"%i",&power_out[row][col]);
           mean += power_out[row][col];
       }

    /*  Calculate the sample mean.  */
    mean = mean/(NROWS*NCOLS);

    /*  Calculate the sample variance.  */
    for (row=0; row<NROWS; row++)
       for (col=0; col<NCOLS; col++)
           var += (power_out[row][col] - mean)*
                  (power_out[row][col] - mean);
    var /= (NROWS*NCOLS-1);

    /*  Print results.  */
    printf("Mean output of power plant, in MegaWatts: %.3f\n",mean);
    printf("Variance of output, in MegaWatts: %.3f\n",var);

    /*  Exit program.  */
    return EXIT_SUCCESS;
}
```

```
/*-------------------------------------------------------------------*/

/*-------------------------------------------------------------------*/
/*  Problem chapter5_17                                            */
/*                                                                 */
/*  This program models a temperature distribution for a grid      */
/*  with six rows and eight columns.  The user enters the          */
/*  temperatures for the four sides.  When a point is updated, its */
/*  new value is used to update the next point. Updating continues,*/
/*  moving across the rows, until the temperature differences of all */
/*  updates are less than a user-entered tolerance value.          */

#include <stdio.h>
#include <stdlib.h>
#include <math.h>
#define NROWS 6
#define NCOLS 8

main()
{
    /*  Declare and initialize variables.  */
    int row, col;
    double t[NROWS][NCOLS], top, right, left, bottom, tolerance,
           update, max_update=0, check;

    /*  Prompt user to enter initial temperatures and tolerance.  */
    printf("Enter initial temperatures (top, right, bottom, left): ");
    scanf("%lf %lf %lf %lf",&top,&right,&bottom,&left);
    printf("Enter tolerance for equilibrium (>0): ");
    scanf("%lf",&tolerance);

    /*  Initialize grid.  */
    for (row=1; row<NROWS-1; row++)
    {
       for (col=1; col<NCOLS-1; col++)
          t[row][col] = 0.0;
       t[row][0] = left;
       t[row][NCOLS-1] = right;
    }
    for (col=0; col<NCOLS; col++)
    {
       t[0][col] = top;
       t[NROWS-1][col] = bottom;
    }

    /*  Update the grid across the rows.  */
    do
    {
       /*  Initialize the maximum update this iteration to zero */
       max_update = 0.0;

       /*  Interior rows */
       for (row=1; row<NROWS-1; row++)
       {
          for(col=1; col<NCOLS-1; col++)
          {
             update = (t[row][col+1] + t[row][col-1] + t[row-1][col]
                      + t[row+1][col])/4;
             check = update - t[row][col];
             if (check > max_update)
```

```
                      max_update = check;
                t[row][col] = update;
            }
     } while(max_update > tolerance);

     printf("Equilibrium values: \n");
     for (row=1; row<NROWS-1; row++)
     {
        for (col=1; col<NCOLS-1; col++)
           printf(" %f.3 ",t[row][col]);
        printf("\n");
     }

     /*  Exit program.  */
     return EXIT_SUCCESS;
}
/*-------------------------------------------------------------*/

/*-------------------------------------------------------------*/
/*  Problem chapter5_18                                        */
/*                                                             */
/*  This program models a temperature distribution for a grid  */
/*  with six rows and eight columns.  The user enters the      */
/*  temperatures for the four sides.  When a point is updated, its */
/*  new value is used to update the next point. Updating continues, */
/*  moving down the columns, until the temperature differences of  */
/*  all updates are less than a user-entered tolerance value.  */

#include <stdio.h>
#include <stdlib.h>
#include <math.h>
#define NROWS 6
#define NCOLS 8

main()
{
     /*  Declare and initialize variables.  */
     int row, col;
     double t[NROWS][NCOLS], top, right, left, bottom, tolerance,
             check, update, max_update=0;

     /*  Prompt user to enter initial temperatures and tolerance.  */
     printf("Enter initial temperatures (top, right, bottom, left): ");
     scanf("%lf %lf %lf %lf",&top,&right,&bottom,&left);
     printf("Enter tolerance for equilibrium: (>0) ");
     scanf("%lf",&tolerance);

     /*  Initialize grid.  */
     for (row=1; row<NROWS-1; row++)
     {
        for (col=1; col<NCOLS-1; col++)
           t[row][col] = 0.0;
        t[row][0] = left;
        t[row][NCOLS-1] = right;
     }
     for (col=0; col<NCOLS; col++)
     {
        t[0][col] = top;
        t[NROWS-1][col] = bottom;
     }
```

74

```
    /*  Update the grid across the rows.  */
    do
    {
        /*  Initialize the maximum update this iteration to zero */
        max_update = 0.0;

        /*  Interior columns */
        for (col=1; col<NCOLS-1; col++)
        {
            for(row=1; row<NROWS-1; row++)
            {
                update = (t[row+1][col] + t[row-1][col] + t[row][col-1]
                        + t[row][col+1])/4;
                check = fabs(update - t[row][col]);
                if (check > max_update)
                   max_update = check;
                t[row][col] = update;
            }
        }
    } while((max_update > tolerance) || (max_update < -tolerance));

    /*  Print results.  */
    printf("Equilibrium values: \n");
    for (row=1; row<NROWS-1; row++)
    {
        for (col=1; col<NCOLS-1; col++)
           printf(" %f.3 ",t[row][col]);
        printf("\n");
    }

    /*  Exit program.  */
    return EXIT_SUCCESS;
}
/*-------------------------------------------------------------------*/

/*-------------------------------------------------------------------*/
/*  Problem chapter5_19                                              */
/*                                                                   */
/*  This program models a temperature distribution for a grid        */
/*  with six rows and eight columns.  The user enters the            */
/*  temperatures for the four sides.  When a point is updated, its   */
/*  old value is used to update adjacent points. Updating continues, */
/*  moving down the columns, until the temperature differences of    */
/*  all updates are less than a user-entered tolerance value.        */

#include <stdio.h>
#include <stdlib.h>
#include <math.h>
#define NROWS 6
#define NCOLS 8

main()
{
    /*  Declare and initialize variables.  */
    int row, col;
    double t[NROWS][NCOLS], top, right, left, bottom, tolerance,
           update[NROWS][NCOLS], max_update=0, check;

    /*  Prompt user to enter initial temperatures and tolerance */
```

```c
      printf("Enter initial temperatures (top, right, bottom, left): ");
      scanf("%lf %lf %lf %lf",&top,&right,&bottom,&left);
      printf("Enter tolerance for equilibrium (>0): ");
      scanf("%lf",&tolerance);

      /*  Initialize grid.  */
      for (row=1; row<NROWS-1; row++)
      {
         for (col=1; col<NCOLS-1; col++)
            t[row][col] = 0.0;
         t[row][0] = left;
         t[row][NCOLS-1] = right;
      }
      for (col=0; col<NCOLS; col++)
      {
         t[0][col] = top;
         t[NROWS-1][col] = bottom;
      }

      /*  Caclulate the updated grid across the rows.  */
      do
      {
         /*  Initialize the maximum update this iteration to zero.  */
         max_update = 0.0;

         /*  Interior rows.  */
         for (row=1; row<NROWS-1; row++)
         {
            for(col=1; col<NCOLS-1; col++)
               update[row][col] = (t[row][col+1] + t[row][col-1] +
                                   t[row-1][col] + t[row+1][col])/4;
         }

         /*  Now check the tolerance.  */
         for (row=0; row<NROWS; row++)
            for (col=0; col<NCOLS; col++)
            {
               check = fabs(update[row][col]-t[row][col]);
               if (check < max_update)
                  max_update = check;
               /*  Set up for next iteration.  */
               t[row][col]=update[row][col];
            }
      } while((max_update > tolerance) || (max_update < -tolerance));

      printf("Equilibrium value is approximately: %f \n",t[0][0]);

      /*  Exit program.  */
      return EXIT_SUCCESS;
}
/*--------------------------------------------------------------------*/

/*--------------------------------------------------------------------*/
/*  Function chapter5_20                                              */
/*                                                                    */
/*  This funtion receives a double array a of size N by N+1, where    */
/*  N is a symbolic constant.  A second parameter is a double         */
/*  array soln of size N.  The function solves the system of          */
/*  equations represented by array a, and returns the solution in     */
/*  the array soln.                                                   */
```

```
void gauss(double a[N][N+1], double soln[N])
{
    /*  Declare variables.  */
    int i, j, k;
    double sum, mult;

    /*  First do the elimination.  */
    for (i=0; i<N-1; i++)
    {
        /*  Can't divide by zero, so can't solve.  */
        if (a[i][i] == 0)
        {
            printf("Can't solve this equation.\n");
            return;
        }
        for (k=1; k<N-i; k++)
        {
            mult = -a[i+k][i] / a[i][i];
            for (j=0; j<N+1; j++)
                a[i+k][j] += mult * a[i][j];
        }
    }

    /*  Now do the backwards substitution.  */
    for (i=N-1; i>=0; i--)
    {
        sum = a[i][N];
        for (k=1; k<N-i; k++)
            sum += -a[i][N-k] * soln[N-k];
        soln[i] =  sum / a[i][N-k];
    }

    /*  Void return.  */
    return;
}
/*------------------------------------------------------------------*/

/*------------------------------------------------------------------*/
/*  Function chapter5_21                                            */
/*                                                                  */
/*  This function that receives a two-dimensional array and a pivot */
/*  value that specifies the coefficient of interest, j.  The       */
/*  function reorders all equations starting with the jth equation  */
/*  such that the jth equation will have the largest coefficient    */
/*  (in absolute value) in the jth position.  Assume that the size  */
/*  of the array is N by N+1 where N is a symbolic constant.        */

void pivot_r(double a[N][N+1], int j)
{
    /*  Declare variables.  */
    int i, row;
    double max_j, tmp;

    /*  Assume that the jth row is already the correct one.  */
    max_j = a[j][j];
    row = j;

    /*  For each equation, starting with the jth.  */
    for (i=j; i<N; i++)
```

```
      /* Find the maximum value in the jth position.  */
      if (fabs(a[i][j]) > max_j)
      {
         max_j = a[i][j];
         row = i;
      }

   /* Now put the row with the maximum absolute value in the jth  */
   /* position in the jth row and the jth row in its place.        */
   if (row != j)
      for (i=0; i<=N; i++)
      {
         tmp = a[j][i];
         a[j][i] = a[row][i];
         a[row][i] = tmp;
      }

   /*  Void return.  */
   return;
}
/*-------------------------------------------------------------------*/

/*-------------------------------------------------------------------*/
/*  Function chapter5_22                                              */
/*                                                                   */
/*  This funtion receives a double array a of size N by N+1, where   */
/*  N is a symbolic constant.  A second parameter is a double        */
/*  array soln of size N.  The function solves the system of         */
/*  equations represented by array a, and returns the solution in    */
/*  the array soln. Row pivoting is performed before each variable   */
/*  is eliminated.                                                   */

void gauss_2(double a[N][N+1], double soln[N])
{
   /* Declare variables and function prototypes */
   int i, j, k;
   double sum, mult;
   void pivot_r(double a[N][N+1], int j);

   /*  First do the elimination.  */
   for (i=0; i<N-1; i++)
   {
      /*  First pivot, then eliminate.  */
      pivot_r(a,i);

      /*  Can't divide by zero.  */
      if (a[i][i] == 0)
      {
         printf("Can not solve this equation.\n");
         return;
      }

      for (k=1; k<N-i; k++)
      {
         mult = -a[i+k][i] / a[i][i];
         for (j=0; j<N+1; j++)
            a[i+k][j] += mult * a[i][j];
      }
   }
```

```
       /*  Now do the backwards substitution.  */
       for (i=N-1; i>=0; i--)
       {
          sum = a[i][N];
          for (k=1; k<N-i; k++)
              sum += -a[i][N-k] * soln[N-k];
          soln[i] =  sum / a[i][N-k];
       }

    /*  Void return.  */
    return;
}
/*-------------------------------------------------------------------/*

/*-------------------------------------------------------------------*/
/*  Function chapter5_23                                            */
/*                                                                  */
/*  This function that receives a two-dimensional array and a pivot */
/*  value that specifies the coefficient of interest, j.  The       */
/*  function reorders all columns starting with the jth column      */
/*  such that the jth column will have the largest coefficient      */
/*  (in absolute value) in the jth position.  Assume that the size  */
/*  of the array is N by N+1 where N is a symbolic constant.        */

void pivot_c(double a[N][N+1], int j, double k[N])
{
    /*  Declare variables.  */
    int i, col;
    double max_j, tmp;

    /*  Assume that the jth column is already the correct one.  */
    max_j = a[j][j];
    col = j;

    /*  For each column, starting with the jth.  */
    for (i=j; i<N; i++)
       /* Find the maximum value of the columns in these equations */
       if (fabs(a[j][i]) > max_j)
       {
          max_j = a[j][i];
          col= i;
       }

    /*  Now put the column with the maximum absolute value in  */
    /*  the jth position in the jth column and the jth column  */
    /*  in its place.                                          */
    if (col != j)
       for (i=0; i<N; i++)
       {
          tmp = a[i][j];
          a[i][j] = a[i][col];
          a[i][col] = tmp;
       }

    /*  Keep track of the switch in k.  */
    tmp = k[j];
    k[j] = k[col];
    k[col] = tmp;

    /*  Return to calling function.  */
```

```
      return;
}
/*------------------------------------------------------------------*/

/*------------------------------------------------------------------*/
/*  Function chapter5_24                                            */
/*                                                                  */
/*  This funtion receives a double array a of size N by N+1, where  */
/*  N is a symbolic constant.  A second parameter is a double       */
/*  array soln of size N.  The function solves the system of        */
/*  equations represented by array a, and returns the solution in   */
/*  the array soln. Column pivoting is performed before each        */
/*  variable is eliminated.                                         */

void gauss_2(double a[N][N+1], double soln[N])
{
   /*  Declare variables and function prototypes.  */
   int i, j, m;
   double k[N], sum, mult, tmp[N];
   void pivot_c(double a[N][N+1], int j, double k[N]);

   /*  Set up k before pivoting.  */
   for (i=0; i<N; i++)
      k[i] = i;

   /*  First do the elimination.  */
   for (i=0; i<N-1; i++)
   {
      /* Column pivot */
      pivot_c(a,i,k);

      /*  Can't divide by zero.  */
      if (a[i][i] == 0)
      {
         printf("Cannot solve this equation.\n");
         return;
      }

      for (m=1; m<N-i; m++)
      {
         mult = -a[i+m][i] / a[i][i];
         for (j=0; j<N+1; j++)
            a[i+m][j] += mult * a[i][j];
      }
   }

   /*  Now do the backwards substitution.  */
   for (i=N-1; i>=0; i--)
   {
      sum = a[i][N];
      for (m=1; m<N-i; m++)
         sum += -a[i][N-m] * tmp[N-m];
      tmp[i] =  sum / a[i][N-m];
   }
   /*  Reorder solutions based on k.  */
   for (i=0; i<N; i++)
      soln[i] = tmp[k[i]];

   /*  Void return.  */
   return;
```

```
}
/*------------------------------------------------------------------*/


/*------------------------------------------------------------------*/
/*  Function chapter5_25                                            */
/*                                                                  */
/*  This funtion receives a double array a of size N by N+1, where  */
/*  N is a symbolic constant.  A second parameter is a double       */
/*  array soln of size N.  The function solves the system of        */
/*  equations represented by array a, and returns the solution in   */
/*  the array soln. Row pivoting and column pivoting are performed  */
/*  before each variable is eliminated.                             */
void gauss_3(double a[N][N+1], double soln[N])
{
   /*  Declare variables and function prototypes.  */
   int i, j, m;
   double k[N], sum, mult, tmp[N];
   void pivot_r(double a[N][N+1], int j);
   void pivot_c(double a[N][N+1], int j, double k[N]);

   /*  Set up k before pivoting.  */
   for (i=0; i<N; i++)
      k[i] = i;

   /*  First do the elimination.  */
   for (i=0; i<N-1; i++)
   {
      /*  Row pivot.  */
      pivot_r(a,i);

      /*  Column pivot.  */
      pivot_c(a,i,k);

      /*  Can't divide by zero.  */
      if (a[i][i] == 0)
      {
         printf("Can not solve this equation.\n");
         return;
      }

      for (m=1; m<N-i; m++)
      {
         mult = -a[i+m][i] / a[i][i];
         for (j=0; j<N+1; j++)
            a[i+m][j] += mult * a[i][j];
      }
   }

   /*  Now do the backwards substitution.  */
   for (i=N-1; i>=0; i--)
   {
      sum = a[i][N];
      for (m=1; m<N-i; m++)
         sum += -a[i][N-m] * tmp[N-m];
      tmp[i] =  sum / a[i][N-m];
   }

   /*  Reorder solutions based on k.  */
   for (i=0; i<N; i++)
```

```
        soln[i] = tmp[k[i]];

    /*  Return to calling function.  */
    return;
}
/*------------------------------------------------------------*/


/*------------------------------------------------------------*/
/*  Function chapter5_26                                      */
/*                                                            */
/*  This function computes the minor of a square matrix       */
/*  with four rows and four columns.                          */

double minor(double a[4][4], int row, int col)
{
    /*  Declare and initialize variables.  */
    int i, j, r=0, c=0;
    double m[3][3], det;

    /*  Determine minor matrix.  */
    for (i=0; i<4; i++)
    {
        /* If the row is not the one that is excluded, check the column */
        if (i != row)
        {
            for (j=0; j<4; j++)
                if (j != col)
                {
                    /*  Neither the row or the column is the one  */
                    /*  excluded, so this is a part of the 3x3.   */
                    m[r][c] = a[i][j];
                    c++;
                }

            /*  Update row counter of the 3x3 and start in column 0.  */
            r++;
            c=0;
        }
    }

    /*  Now calculated determinant m.   */
    det = m[0][0]*m[1][1]*m[2][2] +
          m[0][1]*m[1][2]*m[2][0] +
          m[0][2]*m[1][0]*m[2][1] -
          m[2][0]*m[1][1]*m[0][2] -
          m[2][1]*m[1][2]*m[0][0] -
          m[2][2]*m[1][0]*m[0][1];

    /*  Return determinant.  */
    return det;
}
/*------------------------------------------------------------*/


/*------------------------------------------------------------*/
/*  Function chapter5_27                                      */
/*                                                            */
/*  This function computes the cofactor of a square matrix    */
/*  with four rows and four columns, where the (i,j) cofactor */
/*  is (-1)^(i+j) * minor(a,i,j)                              */
```

```
double cofactor(double a[4][4], int i, int j)
{
    /*  Declare function prototypes.  */
    double minor(double a[4][4], int i, int j);

    /*  Return cofactor.  */
    return pow(-1,(i+j)) * (minor(a,i,j));
}
/*------------------------------------------------------------------*/
              (minor function from Problem 5_27)
/*------------------------------------------------------------------*/

/*------------------------------------------------------------------*/
/*  Function chapter5_28                                            */
/*                                                                  */
/*  This function computes the determinant of a square matrix       */
/*  with four rows and four columns by multiplying each element     */
/*  in a given column by its cofactor and adding the products.      */

double det_c(double a[4][4])
{
    /*  Declare variables and function prototypes.  */
    int i, column=0;
    double sum=0;
    double cofactor(double a[4][4], int i, int j);

    /*  Compute determinant.  */
    for (i=0; i<4; i++)
       sum += a[i][column] * cofactor(a,i,column);

    /*  Return determinant.  */
    return sum;
}
/*------------------------------------------------------------------*/

/*------------------------------------------------------------------*/
/*  Function chapter5_29                                            */
/*                                                                  */
/*  This function computes the determinant of a square matrix       */
/*  with four rows and four columns by multiplying each element     */
/*  in a given row by its cofactor and adding the products.         */

double det_r(double a[4][4])
{
    /* Declare variables and function prototypes */
    int i, row=0;
    double sum=0;
    double cofactor(double a[4][4], int i, int j);

    /*  Comptue determinant.  */
    for (i=0; i<4; i++)
       sum += a[row][i] * cofactor(a,row,i);

    /*  Return determinant.  */
    return sum;
}
/*------------------------------------------------------------------*/

/*------------------------------------------------------------------*/
/*  Function chapter5_30                                            */
```

```
/*                                                                    */
/*   This function computes and returns the value of the correlation  */
/*   coefficient for a set of x,y data.                               */

double corr_coef(double x[N], double y[N], int num_pts)
{
   /*  Declare and initialize variables.   */
   int k;
   double sum=0, sigma_xy;

   /*  Sum up the x(k) and y(k).   */
   for (k=0; k<N; k++)
      sum += x[k]*y[k];

   /*  Calculate numerator and denominator of coefficient.   */
   sum = sum/N - mean(x,num_pts)*mean(y,num_pts);
   sigma_xy = std_dev(x,num_pts)*std_dev(y,num_pts);
   if (sigmaxy == 0)
   {
      printf("Cannot divide by zero. Ignore results.\n");
      return sum;
   }

   /*  Return correlation coefficient.   */
   return sum/sigmaxy;
}
/*-------------------------------------------------------------------*/

/*-------------------------------------------------------------------*/
/*   Problem chapter5_31                                             */
/*                                                                   */
/*   This program determines and prints the correlation cofficient   */
/*   for the altitude and ozone mixing ration values in the data     */
/*   file zone1.dat.                                                 */

#include <stdio.h>
#include <stdlib.h>
#include <math.h>
#include "stat_lib.h"
#define N 10
#define FILENAME "zone1.dat"

main()
{
   /*  Declare the variables and function prototypes.   */
   int i = 0;
   double x[N], y[N];
   double corr_coef(double x[], double y[], int num_pts);
   FILE *ozone;

   /*  Open input file.   */
   ozone = fopen(FILENAME,"r");

   /*  Read data.   */
   while (fscanf(ozone,"%lf %lf",&x[i],&y[i]) == 2)
      i++;

   /*  Get the correlation coefficient.   */
   printf("Correlation coefficient: %f\n",corr_coef(x,y,i));
```

```
      /*  Exit program.  */
      return EXIT_SUCCESS;
}
/*------------------------------------------------------------------*/
              (function corr_coef from Problem 5-30)
/*------------------------------------------------------------------*/


/*------------------------------------------------------------------*/
/*  Function chapter5_32                                            */
/*                                                                  */
/*  This function normalize the values in an array.                 */

void norm_1D(double x[], int num_pts)
{
   /*  Declare variables.  */
   int k;
   double minx, maxx;

   /*  Normalize array.  */
   minx = minimum(x,num_pts);
   maxx = maximum(x,num_pts);
   for (k=0; k<num_pts; k++)
      x[k] = (x[k] - minx)/(maxx - minx);

   /*  Void return.  */
   return;
}
/*------------------------------------------------------------------*/


/*------------------------------------------------------------------*/
/*  Function chapter5_33                                            */
/*                                                                  */
/*  This function normalize the values in a two-dimensional double  */
/*  array assuming that NROWS and NCOLS are symbolic constants that */
/*  specify the size of the array.                                  */

void norm_2D(double x[NROWS][NCOLS])
{
   /*  Declare variables.  */
   int j, k;
   double minx, maxx, check, tmp[NCOLS];

   /*  Initialize max and min.  */
   minx = x[0][0];
   maxx = x[0][0];

   /*  Find the maximum and minimum.  */
   for (j=0; j<NROWS; j++)
   {
      for (k=0; k<NCOLS; k++)
         tmp[k]= x[j][k];

      check = minimum(tmp,NCOLS);
      if (minx > check)
         minx = check;

      check = maximum(tmp,NCOLS);
      if (maxx < check)
         maxx = check;
   }
```

```
    /*  Now normalize the data.                                          */
    /*  Special case, all values the same, so just divide by value.  */
    if (minx == maxx)
        if (minx != 0)
            minx = 0;
        else
        {
            printf("All values are zero, no normalization done.\n");
            return;
        }

    for(j=0; j<NROWS; j++)
        for (k=0; k<NCOLS; k++)
            x[j][k] = (x[j][k] - minx)/(maxx - minx);

    /*  Void return.  */
    return;
}
/*------------------------------------------------------------------*/
```

Chapter 6

```
/*------------------------------------------------------------------*/
/*   Function chapter6_1                                            */
/*                                                                  */
/*   This function converts radius, diameter, and area measurements */
/*   for a circle from units of inches and square inches to units   */
/*   of feet and square feet.                                       */

void convert_ft(double *r, double *d, double *a)
{
    #define IN_PER_FOOT 12
    #define SQ_IN_PER_SQ_FOOT 144
    #define PI 3.1415927

    /*   Perform conversions.   */
    *r = *r/IN_PER_FOOT;
    *d = *d/IN_PER_FOOT;
    *a = *a/SQ_IN_PER_SQ_FOOT;

    /*   Void return.   */
    return;
}
/*------------------------------------------------------------------*/


/*------------------------------------------------------------------*/
/*   Function chapter6_2                                            */
/*                                                                  */
/*   This function reorders the values in three integer variables   */
/*   such that the values are in ascending order. #include <stdio.h>

void reorder(int *a, int *b, int *c)
{
    /*   Declare function prototypes.   */
    void swapit(int *a, int *b);

    /*   Reorder values if needed.   */
    if ((*a<=*b) && (*b<=*c))
       /* Already in order, do nothing. */
       return;
    if (*a > *b)
       swapit(a,b);
    if ( *a > *c)
       swapit(a,c);
    if (*b > *c)
       swapit(b,c);

    /*   Void return.   */
    return;
}
/*------------------------------------------------------------------*/
/* This function swaps two integer values.                          */

void swapit(int *a, int *b)
{
    /*   Declare variables.   */
    int tmp;

    /*   Swap two integer values.   */
```

```c
    tmp = *a;
    *a = *b;
    *b = tmp;

    /*  Void return.  */
    return;
}
/*------------------------------------------------------------------*/

/*------------------------------------------------------------------*/
/*  Function chapter6_3                                           */
/*                                                                */
/*  This functio determines the maximum and minimum values from   */
/*  a one-dimensional integer array.                              */

void ranges(int x[], int npts, int *max_ptr, int *min_ptr)
{
    /*  Declare variables.  */
    int i;

    /*  Initialize pointers.  */
    *min_ptr = x[0];
    *max_ptr = x[0];

    /*  Find maximum and minimum values in array.  */
    for (i=1; i<npts; i++)
    {
        if (x[i] < *min_ptr)
            *min_ptr = x[i];
        if (x[i] > *max_ptr)
            *max_ptr = x[i];
    }

    /*  Void return.  */
    return;
}
/*------------------------------------------------------------------*/

/*------------------------------------------------------------------*/
/*                                                                */
/*  Function chapter6_4                                           */
/*                                                                */
/*  This function returns the double average value of a           */
/*  one-dimensional integer array, in addition to determining the */
/*  number of values in the array that are greater than the average. */
double average(int x[], int npts, int *gtr)
{
    /*  Declare and initialize variables.  */
    int i;
    double ave=0.0;

    /*  Initialize pointer.  */
    *gt r= 0;

    /*  Find average.  */
    for (i=0; i<npts; i++)
        ave += x[i];
    ave /= npts;

    /* Now find number of values greater than average.  */
    for (i = 0; i < npts; i++)
```

```c
        if (x[i] > ave)
            (*gtr)++;

    /*  Return average.  */
    return ave;
}
/*-----------------------------------------------------------------*/

/*-----------------------------------------------------------------*/
/*  Function chapter6_5                                            */
/*                                                                 */
/*  This function returns the number of positive values, zero      */
/*  values, and negative values in an integer array.

void signs(int x[], int npts, int *npos, int *nzero, int *nneg)
{
    /*  Declare variables.  */
    int i;

    /*  Initialize pointers */
    *npos = *nneg = *nzero = 0;

    /*  Count values in each category.  */
    for (i=0; i<npts; i++)
        if (x[i] > 0)
            (*npos)++;
        else
            if (x[i] < 0)
                (*nneg)++;
            else
                (*nzero)++;

    /*  Void return.  */
    return;
}
/*-----------------------------------------------------------------*/

/*-----------------------------------------------------------------*/
/*  Function chapter6_6                                            */
/*                                                                 */
/*  This function fills a vector with zeros.                       */

void zeros(int x[], int n)
{
    /*  Declare variables.  */
    int i;

    /*  Fill vector with zeros.  */
    for (i=0; i<n; i++)
        x[i] = 0;

    /*  Void return.  */
    return;
}
/*-----------------------------------------------------------------*/

/*-----------------------------------------------------------------*/
/*  Function chapter6_7                                            */
/*                                                                 */
/*  This function fills a vector with ones.
```

```
void ones(int x[], int n)
{
   /*  Declare variables.  */
   int i;

   /*  Fill vector with ones.  */
   for (i=0; i<n; i++)
      x[i] = 1;

   /*  Void return.  */
   return;
}
/*-------------------------------------------------------------------*/

/*-------------------------------------------------------------------*/
/*  Function chapter6_8                                            */
/*                                                                 */
/*  This function computes the sum of the values in the array,

int v_sum(int x[], int n)
{
   /*  Declare and initialize variables.  */
   int sum=0, i;

   /*  Compute sum of array values.  */
   for (i=0; i<n; i++)
      sum += x[i];

   /*  Return sum.  */
   return sum;
}
/*-------------------------------------------------------------------*/

/*-------------------------------------------------------------------*/
/*  Function chapter6_9                                            */
/*                                                                 */
/*  This function reverses the order of the values in a vector.    */

void v_rev(int x[], int n)
{
   /*  Declare variables.  */
   int tmp, i, midpoint;

   /*  Reverse values.  */
   midpoint = n/2;
   for (i=0; i<midpoint; i++)
   {
      tmp = x[i];
      x[i] = x[n-i];
      x[n-i] = tmp;
   }

   /*  Void return.  */
   return;
}
/*-------------------------------------------------------------------*/

/*-------------------------------------------------------------------*/
/*  Function chapter6_10                                           */
```

```
/*                                                                     */
/*  This function replaces values in an array with their               */
/*  absolute values.

void v_abs(int x[], int n)
{
   /*  Declare variables.  */
   int i;

   /*  Replace values with absolute values.  */
   for (i=0; i<n; i++)
      x[i] = abs(x[i]);

   /*  Void return.  */
   return;
}
/*---------------------------------------------------------------------*/


/*---------------------------------------------------------------------*/
/*  Function chapter6_11                                                */
/*                                                                     */
/*  This function computes the sum of the ith row in a table            */
/*  containing 10 rows and 8 columns of values.                         */

double row_sum(double table[10][8], int i)
{
   /*  Declare and initialize variables.  */
   int j;
   double sum=0.0;

   /*  Compute sum of ith row.  */
   for (j=0; j<8; j++)
      sum += table[i][j];

   /*  Return sum.  */
   return sum;
}
/*---------------------------------------------------------------------*/


/*---------------------------------------------------------------------*/
/*  Function chapter6_12                                                */
/*                                                                     */
/*  This function computes the sum of the ith column in a table         */
/*  containing 10 rows and 8 columns of values.                         */

double col_sum(double table[10][8], int j)
{
   /*  Declare and initialize variables.  */
   int i;
   double sum=0.0;

   /*  Compute column sum.  */
   for (i=0; i<10; i++)
      sum += table[i][j];

   /*  Return sum.  */
   return sum;
}
/*---------------------------------------------------------------------*/
```

Chapter 7

```
/*-----------------------------------------------------------------*/
/*   Problem chapter7_1                                            */
/*                                                                 */
/*   This program reads a data file that should contain only integer */
/*   values, and thus should contain only digits, plus or minus    */
/*   signs, and white space.  The program should print a count of  */
/*   any invalid characters located.                               */

#include <stdio.h>
#include <stdlib.h>
#include <ctype.h>
#define FILENAME "data.dat"
#define NEWLINE '\n'
#define TAB '\t'

main()
{
   /*  Declare variables.  */
   int counter=0;
   char c;
   FILE *file1;

   /*  Open input file.  */
   file1 = fopen(FILENAME, "r");

   /*  Check file characters to see if legal.  */
   /*  If not legal, count it.                 */
   do
   {
      c = fgetc(file1);
      switch (c)
      {
         case '+':
            break;
         case '-':
            break;
         case NEWLINE:
            break;
         case TAB:
            break;
         case ' ':
            break;
         case EOF:
            break;
         default:
            if (!(isdigit(c)))
               counter++;
            break;
      }

   } while (c != EOF);

   /*  Print results.  */
   printf("Number of invalid characters is: %i\n", counter);

   /*  Close file and exit program.  */
   fclose(file1);
```

```c
      return EXIT_SUCCESS;
}
/*-------------------------------------------------------------------*/

/*-------------------------------------------------------------------*/
/*  Problem chapter7_2                                               */
/*                                                                   */
/*                                                                   */
/*  This program analyzes a data file that has been determined      */
/*  to contain only integer values and white space.  The program    */
/*  prints the number of lines in the file and the number of integer */
/*  values (not integer digits).                                    */

#include <stdio.h>
#include <stdlib.h>
#include <ctype.h>
#define FILENAME "data.dat"
#define NEWLINE '\n'
#define TAB '\t'
#define TRUE 1
#define FALSE 0

main()
{
   /*  Declare and initialize variables.  */
   int white=TRUE, int_count=0, line_count=0;
   char c;
   FILE *file1;

   /*  Open input file.  */
   file1 = fopen(FILENAME, "r");

   /*  Look for characters that are not integers and white space.  */
   do
   {
      c = fgetc(file1);
      switch (c)
      {
         case NEWLINE:
            line_count++;
            /* fall through */
         case ' ':
         case EOF:
            if (white==FALSE)
            {
               /* Previous character was a digit, and now the */
               /* integer is complete, so count it. */
               int_count++;
               white = TRUE;
            }
            break;
         default:
            /* Digit--but don't count until new white space. */
            white = FALSE;
            break;
      }
   } while (c != EOF);

   /*  Print results.  */
   printf("Number of integers: %i.  Number of lines: %i.\n",
```

93

```
                int_count,line_count);

     /*  Close file and exit program.  */
     fclose(file1);
     return EXIT_SUCCESS;
}
/*------------------------------------------------------------------*/


/*------------------------------------------------------------------*/
/*  Problem chapter7_3                                              */
/*                                                                  */
/*  This program reads a file that contains only integers, but      */
/*  some of the integers have embedded commas, as in 145,020.  The  */
/*  program copies the information to a new file, removing any       */
/*  commas from the information.                                     */

#include <stdio.h>
#include <stdlib.h>
#include <ctype.h>
#define INNAME "in.dat"
#define OUTNAME "out.dat"
#define NEWLINE '\n'
#define TAB '\t'
#define TRUE 1
#define FALSE 0

main()
{
     /*  Declare variables.  */
     char c;
     FILE *newfile, *oldfile;

     /*  Open the files for reading and writing respectively.  */
     oldfile = fopen(INNAME, "r");
     newfile = fopen(OUTNAME, "w");

     /*  Look for commas, and write the data to the new file.  */
     do
     {
        c = fgetc(oldfile);
        if ( c != ',')
           putc(c,newfile);
     } while (c != EOF);

     /*  Close files and exit program.  */
     fclose(newfile);
     fclose(oldfile);
     return EXIT_SUCCESS;
}
/*------------------------------------------------------------------*/


/*------------------------------------------------------------------*/
/*  Problem chapter7_4                                              */
/*                                                                  */
/*  This program reads a file containing integer and floating-point */
/*  values seperated by commas, which may or may not be followed     */
/*  by additional white space. A new file is generated that contains */
/*  the integers and floating-point values seperated only by a       */
/*  single space between the values.                                 */
```

```
#include <stdio.h>
#include <stdlib.h>
#include <ctype.h>
#define INNAME "in.dat"
#define OUTNAME "out.dat"
#define NEWLINE '\n'
#define TAB '\t'

main()
{
   /*  Declare variables.  */
   char c;
   FILE *newfile, *oldfile;

   /*  Open the files for reading and writing respectively.  */
   oldfile = fopen(INNAME, "r");
   newfile = fopen(OUTNAME, "w");

   /*  Put only a single space between numerical values.  */
   do
   {
      c = fgetc(oldfile);
      switch (c)
      {
         case TAB:
         case ' ':
            break;
         case ',':
            /*  Check to see if a space was already written.  */
            fputc(' ',newfile);
            break;
         default:
            /*  Either a digit or a new line or EOF.  */
            fputc(c,newfile);
            break;
      }
   } while (c != EOF);

   /*  Close files and exit program.  */
   fclose(newfile);
   fclose(oldfile);
   return EXIT_SUCCESS;
}
/*-------------------------------------------------------------------*/

/*-------------------------------------------------------------------*/
/*  Problem chapter7_5                                             */
/*                                                                */
/*  This program reads a file containing data values computed by   */
/*  an accounting software package.  While the file contains only  */
/*  numerical information, the values may contain embedded commas  */
/*  and dollar signs.  The program generates a new file that       */
/*  contains the values with the commas and dollar signs removed,  */
/*  and with a leading minus sign instead of parantheses.          */

#include <stdio.h>
#include <stdlib.h>
#include <ctype.h>
#define INNAME "in.dat"
#define OUTNAME "out.dat"
```

```
#define NEWLINE '\n'
#define TAB '\t'

main()
{
    /*  Declare variables.  */
    char c;
    FILE *newfile, *oldfile;

    /*  Open files for reading and writing respectively.  */
    oldfile = fopen(INNAME, "r");
    newfile = fopen(OUTNAME, "w");

    /*  Now change the format.  */
    do
  {
        c = fgetc(oldfile);
        switch (c)
        {
            case ')':
            case '$':
            case ',':
                break;
            case '(':
                c = '-';
                /* fall thru */
            default:
                fputc(c, newfile);
                break;
        }
    } while (c != EOF);

    /*  Close files and exit program.  */
    fclose(newfile);
    fclose(oldfile);
    return EXIT_SUCCESS;
}
/*----------------------------------------------------------------------*/

/*----------------------------------------------------------------------*/
/*   Problem chapter7_6                                                 */
/*                                                                      */
/*   This program compares two files.  The program should print a       */
/*   message indicating that the files are exactly the same, or that    */
/*   there are differences.  If the fields are different, the program   */
/*   should print the line numbers for lines that are not the same.     */

#include <stdio.h>
#include <stdlib.h>
#include <ctype.h>
#define AFILE "decoded.dat"
#define BFILE "in.dat"
#define NEWLINE '\n'
#define TAB '\t'

main()
{
    /*  Declare and initialize variables.  */
    char a,b;
    int line=1, diff=0, line_flag=0;
```

96

```c
   FILE *afile, *bfile;

   /*  Open input files.  */
   afile = fopen(AFILE, "r");
   bfile = fopen(BFILE, "r");

   /*  Now compare the values.  */
   do
   {
      /*  Get the character from each file.  */
      a = fgetc(afile);
      b = fgetc(bfile);

      /*  If they are not the same, increment a count of differences  */
      /*  and note that there is a difference on this line.           */
      if (a != b)
      {
         diff++;
         line_flag = 1;
      }

      /*  If we are the end of a line in the a-file      */
      /*  print a message if there was an error on this line   */
      /*  and reset the flag indicating an error on the line.  */
      if (a == NEWLINE)
      {
         if (line_flag)
         {
            printf("Files differ on line %i\n", line);
            line_flag = 0;
         }

         /* The next character will be on the next line */
         line++;
      }
   } while ((a!=EOF) && (b!=EOF));

   /*  Check that files are the same length.  */
   if (( a!=EOF ) || (b!=EOF))
      printf("Files differ in length\n");
   else
      /*  Check to see that there are no differences.  */
      if (diff == 0)
   printf("Files are the same\n");

   /*  Close files and exit prgram.  */
   fclose(afile);
   fclose(bfile);
   return EXIT_SUCCESS;
}
/*-------------------------------------------------------------------*/


/*-------------------------------------------------------------------*/
/*  Function chapter7_7                                              */
/*                                                                  */
/*  This function receives an integer array.  If all the values are  */
/*  between 0 and 50, it prints a bar graph and returns a value of   */
/*  0; otherwise, it does not print a bar graph, and returns a       */
/*  value of 1.                                                      */
```

97

```
int bargraph_1(int count, int data[])
{
   /*  Declare variables.  */
   int i, splats;

   /*  Return failure if data not in range.  */
   for (i=0; i<count; i++)
      if ((data[i] < MIN ) || (data[i] > MAX))
        return 1;

   /*  The data is okay, so draw the graph with the splat character.  */
   for (i=0; i<count; i++)
   {
      printf("%2i\t", data[i]);
      for (splats = 0; splats < data[i]; splats++)
        printf("*");
      printf("\n");
   }

   /*  Return success.  */
   return 0;
}
/*------------------------------------------------------------------*/

/*------------------------------------------------------------------*/
/*  Function chapter7_8                                           */
/*                                                                */
/*  This function receives an integer array.  If all the values are  */
/*  between 0 and 50, it prints a bar graph and returns a value of   */
/*  0; otherwise, it does not print a bar graph, and returns a       */
/*  value of 1. It prints two lines of asterisks for each bar.       */

int bargraph_2(int count, int data[])
{
   /*  Declare variables and function prototypes.  */
   int i;
   void print_splat(int);

   /*  Return failure if data not in range.  */
   for (i=0; i<count; i++)
      if ((data[i] < MIN ) || (data[i] > MAX))
        return 1;

   /*  The data is okay, so draw the graph with the splat character.  */
   for (i=0; i<count; i++)
   {
      printf("%2i\t", data[i]);
      print_splat(data[i]);
      printf("\t");
      print_splat(data[i]);
   }

   /*  Return success.  */
   return 0;
}
/*------------------------------------------------------------------*/
/*  This function prints the requiste number of asterisks (splats).  */

void print_splat(int total)
{
```

```c
   /*  Declare variables.  */
   int splats;

   /*  Print asterisks.  */
   for (splats = 0; splats < total; splats++)
      printf("*");
   printf("\n");

   /*  Void return.  */
   return;
}
/*------------------------------------------------------------------*/


/*------------------------------------------------------------------*/
/*  Function chapter7_9                                             */
/*                                                                  */
/*  This function receives an integer array.  If all the values are */
/*  between 0 and 50, it prints a bar graph and returns a value of  */
/*  0; otherwise, it does not print a bar graph, and returns a      */
/*  value of 1.  For each bar, it print asterisks for any part of   */
/*  the bar up to the average value, and then it prints plus signs  */
/*  for any part of the bar that is over the average.               */

int bargraph_3(int count, int data[])
{
   /*  Declare and initialize variables.  */
   int i, position, ave=0;

   /*  Return failure if data not in range.  */
   for (i=0; i<count; i++)
      if ((data[i] < MIN ) || (data[i] > MAX))
         return 1;
      else
         ave += data[i];
   ave = (int) ave/count;

   /*  The data is okay, so draw the graph with the splat character.  */
   for (i=0; i<count; i++)
   {
      printf("%2i\t", data[i]);
      for (position = 0; position < data[i]; position++)
         if (position < ave)
            printf("*");
         else
            printf("+");
      printf("\n");
   }

   /*  Return success.  */
   return 0;
}
/*------------------------------------------------------------------*/


/*------------------------------------------------------------------*/
/*  Function chapter7_10                                            */
/*                                                                  */
/*  This function generates a bar graph that corresponds to         */
/*  scaled integers between 0 and 50 that are computed from integers */
/*  in an array.  The bargraph also contains the original values    */
/*  next to the bars.                                               */
```

```c
void bar_4(int count, int data[])
{
   /*  Declare and initialize variables.  */
   int i, position, divisor, mindata, maxdata, scale=1;

   /*  Initialize mindata and maxdata.  */
   mindata = data[0];
   maxdata = data[0];

   /*  Get the data range.  */
   for (i=0; i<count; i++)
      if (data[i] > maxdata)
         maxdata = data[i];
      else
         if (data[i] < mindata)
            mindata = data[i];

   /*  Calculate divisor.  */
   divisor = maxdata - mindata;
   if (divisor == 0)
      /* All values are the same. Scale appropriately.*/
      if (mindata == 0)
         divisor = 1;
      else divisor = mindata;

   /*  Scale the data and print the graph.  */
   for (i=0; i<count; i++)
   {
      printf("%2i\t", data[i]);
      scale = (int) (MAX * (data[i] - mindata)/divisor);
      for (position = 0; position < scale; position++)
         printf("*");
      printf("\n");
   }

   /*  Return success.  */
   return;
}
/*-------------------------------------------------------------------*/

/*-------------------------------------------------------------------*/
/*  Function chapter7_11                                          */
/*                                                               */
/*  This function generates a bar graph that corresponds to      */
/*  scaled integers between 0 and an maximum value that are      */
/*  computed from integers in an array.  The bargraph also       */
/*  contains the original values next to the bars.               */

void bar_5(int count, int data[], int max_bar)
{
   /*  Declare and initialize variables.  */
   int i, position, divisor, mindata, maxdata, scale=1;

   /*  Initialize mindata and maxdata.  */
   mindata = data[0];
   maxdata = data[0];

   /*  Get the data range.  */
   for (i=0; i<count; i++)
```

100

```c
        if (data[i] > maxdata)
           maxdata = data[i];
        else
           if (data[i] < mindata)
              mindata = data[i];

     /*  Calculate divisor.   */
     divisor = maxdata - mindata;
     if (divisor == 0)
        /* All values are the same. Scale appropriately.*/
        if (mindata == 0)
           divisor = 1;
        else divisor = mindata;

     /*  Scale the data and print the graph.   */
     for (i=0; i<count; i++)
     {
        printf("%2i\t", data[i]);
        scale = (int) (max_bar * (data[i] - mindata)/divisor);
        for (position = 0; position < scale; position++)
           printf("*");
        printf("\n");
     }

     /*  Exit program.   */
     return;
}
/*-------------------------------------------------------------------*/

/*-------------------------------------------------------------------*/
/*  Problem chapter7_12                                              */
/*                                                                   */
/*  This program reads the text in a file, then generates a new      */
/*  file that contains the coded text in which characters are        */
/*  replaced by characters using this scheme.  Do not change         */
/*  the newline characters or EOF characters.                        */

#include <stdio.h>
#include <stdlib.h>
#include <ctype.h>
#define FILENAME "in.dat"
#define CRYPTFILE "out.dat"
#define NEWLINE '\n'

main()
{
   /*  Declare variables.   */
   int code;
   char oldchar;
   FILE *original, *encrypted;

   /*  Open files for reading and writing respectively.   */
   original = fopen(FILENAME, "r");
   encrypted = fopen(CRYPTFILE, "w");

   /*  Generate coded message.   */
   do
   {
      oldchar = fgetc(original);
      switch (oldchar)
```

```
            {
               case EOF:
                  break;
               case NEWLINE:
                  fputc(oldchar,encrypted);
                  break;
               default:
                  code = (int) oldchar + 2;
                  fputc((char)code ,encrypted);
                  break;
            }
       } while (oldchar != EOF);

       /*  Exit program.  */
       return EXIT_SUCCESS;
}
/*------------------------------------------------------------------*/

/*------------------------------------------------------------------*/
/*  Problem chapter7_13                                           */
/*                                                                */
/*  This program reads the text in a file that has been encoded by */
/*  a scheme in which characters are replaced by characters two    */
/*  characters to the right in the collating sequence.             */

#include <stdio.h>
#include <stdlib.h>
#include <ctype.h>
#define DECODEDFILE "decoded.dat"
#define CRYPTFILE "out.dat"
#define NEWLINE '\n'

main()
{
       /*  Declare variables.  */
       int decode;
       char oldchar;
       FILE *decoded, *encrypted;

       /*  Open files for reading and writing respectively.  */
       decoded = fopen(DECODEDFILE, "w");
       encrypted = fopen(CRYPTFILE, "r");

       /*  Generate decoded message.  */
       do
       {
          oldchar = fgetc(encrypted);
          switch (oldchar)
          {
             case EOF:
                break;
             case NEWLINE:
                fputc(oldchar,decoded);
                break;
             default:
                decode = (int) oldchar - 2;
                fputc((char) decode, decoded);
                break;
          }
```

```
    } while (oldchar != EOF);

    /*  Exit program.  */
    return EXIT_SUCCESS;
}
/*------------------------------------------------------------------*/


/*------------------------------------------------------------------*/
/*  Problem chapter7_14                                             */
/*                                                                  */
/*  This program reads a data file and determines the number of    */
/*  occurances of each character in the file.  It then prints the   */
/*  characters and the number of times that they occurred.         */

#include <stdio.h>
#include <stdlib.h>
#include <ctype.h>
#define MAX_ASC 127
#define MIN_ASC 0
#define NUM_ASC 128
#define CODEFILE "in.dat"

main()
{
    /*  Declare variables.  */
    int occur[NUM_ASC], i;
    char c;
    FILE *encrypted;

    /*  Open input file.  */
    encrypted = fopen(CODEFILE, "r");

    /*  Initialize array.  */
    for (i = 0; i<NUM_ASC; i++)
      occur[i] = 0;

    /*  Count characters  */
    do
    {
      c = fgetc(encrypted);
      occur[(int) c] ++;
    } while (c != EOF);

    /*  Print results.  */
    for (i = 0; i<NUM_ASC; i++)
      if (occur[i] != 0)
         printf("ACSII:'%c'\t, VALUE: %i\t, OCCURANCES: %i\n",
                (char)i,i,occur[i]);

    /*  Exit program.  */
    return EXIT_SUCCESS;
}
/*------------------------------------------------------------------*/


/*------------------------------------------------------------------*/
/*  Problem chapter7_15                                             */
/*                                                                  */
/*  This program reads a data file and determines the secret        */
/*  message stored by the sequence of first letters of the words.   */
```

```c
#include <stdio.h>
#include <stdlib.h>
#include <ctype.h>
#define CODEFILE "in.dat"
#define TRUE  1
#define FALSE 0
#define NEWLINE '\n'
#define SPACE ' '
#define TAB '\t'

main()
{
   /*  Declare variables.  */
   int inword = FALSE;
   char c;
   FILE *encrypted;

   /*  Open input file.  */
   encrypted = fopen(CODEFILE, "r");

   /*  Determine message.  */
   do
   {
      /*  if not already in a word, this is the first letter.  */
      c = fgetc(encrypted);
      if (!inword)
      {
         printf("%c", c);
         inword = TRUE;
      }
      switch (c)
      {
         case NEWLINE:
            printf("%c",c);
         case TAB:
         case SPACE:
            inword = FALSE;
         default:
            break;
      }
   } while (c != EOF);

   /*  Exit program.  */
   return EXIT_SUCCESS;
}
/*------------------------------------------------------------------*/

/*------------------------------------------------------------------*/
/*  Problem chapter7_16                                           */
/*                                                                */
/*  This program reads a data file and determines the secret      */
/*  message stored in the second letter of each word.             */

#include <stdio.h>
#include <stdlib.h>
#include <ctype.h>
#define CODEFILE "in.dat"
#define TRUE  1
#define FALSE 0
#define NEWLINE '\n'
```

```
#define SPACE ' '
#define TAB '\t'

main()
{
    /*  Declare and initialize variables.  */
    int was_first=FALSE, inword=FALSE;
    char c;
    FILE *encrypted;

    /*  Open input file.  */
    encrypted = fopen(CODEFILE,"r");

    /*  Determine secret message.  */
    do
    {
        /*  If not already in a word, this is the second letter.  */
        c = fgetc(encrypted);
        if (!inword)
        {
            if (was_first)
            {
                printf("%c", c);
                inword = TRUE;
                was_first = FALSE;
            }
            else was_first = TRUE;
        }
        switch (c)
        {
            case NEWLINE:
                printf("%c", c);
            case TAB:
            case SPACE:
                inword = FALSE;
                was_first = FALSE;
            default:
                break;
        }
    } while (c != EOF);


    /*  Exit program.  */
    return EXIT_SUCCESS;
}
/*-------------------------------------------------------------------*/

/*-------------------------------------------------------------------*/
/*  Problem chapter7_17                                              */
/*                                                                   */
/*  This program reads a data file and determines the secret        */
/*  message stored by characters that are three characters to the   */
/*  right in the collating sequence from the first letters of the   */
/*  words in the data file.

#include <stdio.h>
#include <stdlib.h>
#include <ctype.h>
#define CODEFILE "in.dat"
#define TRUE  1
```

```
#define FALSE 0
#define NEWLINE '\n'
#define SPACE ' '
#define TAB '\t'

main()
{
    /* Declare and initialize variables */
    int code, inword=FALSE;
    char c;
    FILE *encrypted;

    /*  Open input file.  */
    encrypted = fopen(CODEFILE, "r");

    /*  Determine secret message.  */
    do
    {
        /*  If not already in a word, this is the first letter.  */
        c = fgetc(encrypted);
        if (!inword)
        {
            code =  (int) c + 3;
            printf("%c", (char)code);
            inword = TRUE;
        }
        switch (c)
        {
            case NEWLINE:
                printf("%c", c);
            case TAB:
            case SPACE:
                inword = FALSE;
            default:
                break;
        }
    } while (c != EOF);


    /*  Exit program.  */
    return EXIT_SUCCESS;
}
/*-------------------------------------------------------------------*/

/*-------------------------------------------------------------------*/
/*  Problem chapter7_18                                           */
/*                                                                */
/*  This program encodes the text in a data file using an integer */
/*  array named key that contains 26 characters.  This key is read */
/*  from the keyboard; the first letter contains the character that */
/*  is to replace the letter a in the data file, the second letter */
/*  contains the letter that is to replace the letter b in the data */
/*  file, and so on.  Assume that all the punctuation is to be     */
/*  replaced by spaces.                                           */

#include <stdio.h>
#include <stdlib.h>
#include <ctype.h>
#define CODEFILE "out.dat"
#define OLDFILE "in.dat"
```

```c
#define TRUE  1
#define FALSE 0
#define NEWLINE '\n'
#define SPACE ' '
#define TAB '\t'
#define BACKSPACE '\b'
#define NUM_LET 26

main()
{
    /*  Declare and initialize variables.  */
    int key[NUM_LET], i=0, j;
    char c;
    FILE *encrypted, *original;

    /*  Get key from user and validate it.  */
    printf("Enter the 26 unique values for the lowercase letters.\n ");
    printf("White space characters are ignored.\n");
    do
    {
        /* Get the key */
        c = getchar();
        switch (c)
        {
            /* Ignore white space characters */
            case SPACE:
            case BACKSPACE:
            case TAB:
            case NEWLINE:
                break;
            default:
                /* Insure unique decodability */
                if ((c<'!') || (c>'~'))
                    printf("Character out of range, try again.\n");
                else
                    /* A legitimate character for the code */
                    {
                        key[i]= (int) c;
                        printf("key[%i] = %c\n", i, key[i]);
                        i++;
                    }
                break;
        }
    } while (i < NUM_LET);

    /*  Now, make sure that the code is unique.  */
    for (i = 1; i<NUM_LET; i++)
    {
        for (j=0; j<i ; j++)
            if (key[i] == key[j])
            {
                printf("Key[%i] is not unique.\n",i);
                return EXIT_FAILURE;
            }
    }

    /*  Open files for reading and writing.  */
    encrypted = fopen(CODEFILE, "w");
    original = fopen(OLDFILE, "r");
```

```c
    /*  Now code the file.  */
    do
    {
       c = fgetc(original);
       if ((c >='a') && (c <= 'z'))
          fputc(key[c-'a'],encrypted);
       else
          switch(c)
          {
             case NEWLINE:
             case SPACE:
             case TAB:
                fputc(c, encrypted);
                break;
             case '.':
             case ',':
             case '?':
             case '!':
             case ':':
             case ';':
                fputc(SPACE, encrypted);
                break;
             default:
                fputc(c, encrypted);
                break;
          }
    } while (c != EOF);

    /*  Exit program.  */
    return EXIT_SUCCESS;
}
/*-------------------------------------------------------------------*/


/*-------------------------------------------------------------------*/
/*   Problem chapter7_19                                             */
/*                                                                   */
/*   This program decodes the text in a data file that has been
/*   encoded using an integer array named key that contains 26
/*   characters.  This key is read from the keyboard; the first
/*   charcter of the key was used to replace the letter a in
/*   the data file, the second character of the key was used to
/*   replace the letter b in the data file, and so on.  All the
/*   punctuation was replaced by spaces.

#include <stdio.h>
#include <stdlib.h>
#include <ctype.h>
#define CODEFILE "out.dat"
#define DECODEFILE "decoded.dat"
#define TRUE   1
#define FALSE 0
#define NEWLINE '\n'
#define SPACE ' '
#define TAB '\t'
#define BACKSPACE '\b'
#define NUM_LET 26

main()
{
    /*  Declare and initialize variables.  */
```

108

```c
int key[NUM_LET], i=0, j;
char c;
FILE *encrypted, *decoded;

/*  Get key and validate it.  */
printf("Enter the 26 unique values for the lowercase letters.\n ");
printf("White space characters are ignored. Use only lowercase\n");
printf("input for most unique decodeability.\n");
do
{
    /* Get the key */
    c = getchar();
    switch (c)
    {
        /* Ignore white space characters */
        case SPACE:
        case BACKSPACE:
        case TAB:
        case NEWLINE:
            break;
        default:
            /* Insure unique decodability */
            if ((c<'!') || (c>'~'))
                printf("Character out of range, try again.\n");
            else
                /* A legitimate character for the code */
                {
                    key[i]= (int) c;
                    printf("key[%i] = %c\n", i, key[i]);
                    i++;
                }
            break;
    }
} while (i < NUM_LET);

/*  Now, make sure that the code is unique.  */
for (i = 1; i<NUM_LET; i++)
{
    for (j=0; j<i ; j++)
        if (key[i] == key[j])
        {
            printf("Key[%i] is not unique.\n",i);
            return EXIT_FAILURE;
        }
}

/*  Open files for reading and writing. */
encrypted = fopen(CODEFILE, "r");
decoded = fopen(DECODEFILE, "w");

/*  Now code the file.  */
do
{
    c = fgetc(encrypted);

    for (i=0; i < NUM_LET; i++)
        if (c == (char)key[i])
        {
            c = (char) (i + (int) 'a');
            break;
```

```
            }

        switch(c)
        {
            case NEWLINE:
            case SPACE:
            case TAB:
                fputc(c, decoded);
                break;
            default:
                fputc(c, decoded);
                break;
        }
    } while (c != EOF);

    /*  Exit program.  */
    return EXIT_SUCCESS;
}
/*------------------------------------------------------------------*/


/*------------------------------------------------------------------*/
/*  Function chapter7_20                                            */
/*                                                                  */
/*  This function prints the information in the arrays in a         */
/*  table with a title and column headings.                        */

void print_elts(int n_elts, int atomic_num[],
                char atomic_sym[103][3], double atomic_wt[])
{
    /*  Declare variables.  */
    int i;

    /*  Print information.  */
    printf("Table of Atomic Element Information\n\n");
    printf("Atomic Number\tAtomic Symbol\tAtomic Weight\n");
    printf("-----------------------------------------\n");
    for (i=0; i<n_elts; i++)
        printf("%i\t\t%s\t\t%7.4f\n",atomic_num[i],
                atomic_sym[i], atomic_wt[i]);

    /*  Void return.  */
    return;
}
/*------------------------------------------------------------------*/


/*------------------------------------------------------------------*/
/*  Function chapter7_21                                            */
/*                                                                  */
/*  This function checks each atomic symbol to be sure that         */
/*  the first letter is uppercase, and that the remaining           */
/*  letters are lowercase.  If a letter is found with the           */
/*  wrong case, the function replaces it with the correct case.     */

void check_symbol(int n_elts, char atomic_symbol[103][3])
{
    /*  Declare variables.  */
    int i;

    /*  Check each character.  */
```

```c
   for (i = 0; i < n_elts; i++)
   {
      if (!(isalpha(atomic_symbol[i][0])) ||
          !(isalpha(atomic_symbol[i][1])))
      {
         printf("Warning:  Error in table row %i\n", i);
         return;
      }
      if (islower(atomic_symbol[i][0]))
         atomic_symbol[i][0] = toupper(atomic_symbol[i][0]);
      if (isupper(atomic_symbol[i][1] ))
         atomic_symbol[i][1] = tolower(atomic_symbol[i][1]);
   }

   /*  Void return.  */
   return;
}
/*-------------------------------------------------------------------*/

/*-------------------------------------------------------------------*/
/*  Function chapter7_22                                             */
/*                                                                   */
/*  This function prints the atomic symbol and the atomic weight     */
/*  for each element whose weight is above a specified weight.       */
/*  The function returns the number of symbols printed.              */

int select_wt(int n_elts, char atomic_sym[103][3],
              double atomic_wt[], double limit)
{
   /*  Declare and initialize variables.  */
   int i, total = 0;

   /*  Print header. */
   printf("Table of Atomic Element Information\n\n");
   printf("\tAtomic Symbol\tAtomic Weight\n");
   printf("-----------------------------------------\n");

   /*  Print the symbols and weights which meet criteria.  */
   for (i=0; i<n_elts; i++)
      if (atomic_wt[i] > limit)
      {
         total++;
         printf("\t%s\t\t%7.4f\n", atomic_sym[i], atomic_wt[i]);
      }

   /*  Return the number of symbols selected. */
   if (total > 0)
      return total;
   else
      printf("No elements met the criteria\n");
      return 0;
}
/*-------------------------------------------------------------------*/

/*-------------------------------------------------------------------*/
/*  Function chapter7_23                                             */
/*                                                                   */
/*  This function prints the atomic symbol and the atomic weight     */
/*  for each element whose weight is between specified lower          */
/*  weight and upper weight.  The function returns the number         */
```

111

```
/*  of symbols printed.

int limit_wt(double lower, double upper,int n_elts,
             char atomic_sym[103][3], double atomic_wt[])
{
   /*  Declare and initialize variables.  */
   int i, total = 0;

   /*  Print header.  */
   printf("Table of Atomic Element Information\n\n");
   printf("\tAtomic Symbol\tAtomic Weight\n");
   printf("---------------------------------------\n");

   /*  Print the symbols and weights which meet criteria.  */
   for (i=0; i<n_elts; i++)
      if ((atomic_wt[i] > lower) && (atomic_wt[i] < upper))
      {
         total++;
         printf("\t%s\t\t%7.4f\n", atomic_sym[i], atomic_wt[i]);
      }

   /*  Return the number of symbols selected.  */
   if (total > 0)
      return total;
   else
      printf("No elements met the criteria\n");
      return 0;
}
/*-------------------------------------------------------------------*/

/*-------------------------------------------------------------------*/
/*  Function chapter7_24                                             */
/*                                                                   */
/*  This function copies the atomic symbols to another array,        */
/*  sorts the new array of symbols, and prints an alphabetical       */
/*  listing of the symbols.                                          */

void sort_sym(int n_elts, char atomic_sym[103][3])
{
   /*  Declare variables.  */
   int i, j;
   char alpha_sym[103][3], tmp[3];

   /*  Print header.  */
   printf("\tAlphabetized Atomic Symbols\n");
   printf("\t---------------------------\n");

   /*  Copy and sort symbols.  */
   for (i=0; i<n_elts; i++)
      strncpy(alpha_sym[i], atomic_sym[i], 3);
   for (i=0; i<n_elts; i++)
      for (j=0 ; j<n_elts; j++)
      {
         if (strncmp(alpha_sym[i], alpha_sym[j], 3) <0)
         {
            strncpy(tmp,alpha_sym[i],3);
            strncpy(alpha_sym[i], alpha_sym[j],3);
            strncpy(alpha_sym[j],tmp,3);
         }
      }
```

```
      /*  Print symbols.  */
      for (i=0; i<n_elts; i++)
         printf("\t\t%s\n", alpha_sym[i]);

      /*  Void return.  */
      return;
}
/*------------------------------------------------------------------*/

/*------------------------------------------------------------------*/
/*  Function chapter7_25                                            */
/*                                                                  */
/*  This function reorders the values in the three arrays           */
/*  such that the symbols are alphabetical, and such that           */
/*  the order of the values in the other two arrays is              */
/*  changed to correspond to the new order of the symbols.          */

void reorder(int n_elts, int atomic_num[],
             char atomic_sym[103][3], double atomic_wt[])
{
      /*  Declare variables.  */
      int i, j;
      char tmpc[3], tmpd;

      /*  Print header.  */
      printf("Table of Atomic Element Information, Alphabetized\n\n");
      printf("\tAtomic Number\tAtomic Symbol\tAtomic Weight\n");
      printf("\t-------------------------------------------\n");

      /*  Sort data.  */
      for (i=0; i<n_elts; i++)
         for (j=0 ; j< n_elts; j++)
         {
            if (strncmp(atomic_sym[i],atomic_sym[j],3) < 0)
            {
               strncpy(tmpc,atomic_sym[i],3);
               strncpy(atomic_sym[i], atomic_sym[j],3);
               strncpy(atomic_sym[j],tmpc,3);

               tmpd = atomic_num[i];
               atomic_num[i] = atomic_num[j];
               atomic_num[j] = (int) tmpd;

               tmpd = atomic_wt[i];
               atomic_wt[i] = atomic_wt[j];
               atomic_wt[j] = tmpd;
            }
         }

      /*  Print results.  */
      for (i=0; i<n_elts; i++)
         printf("\t\t%i\t\t%s\t%f\n",
                atomic_num[i],atomic_sym[i],atomic_wt[i]);

      /*  Void return.  */
      return;
}
/*------------------------------------------------------------------*/
```

```
/*------------------------------------------------------------------*/
/*  Function chapter7_26                                            */
/*                                                                  */
/*  This function converts an integer numerical value to an         */
/*  integer array that contains the characters of the              */
/*  original integer.                                              */

int int_to_char(int x, int ch[], int max_ch)
{
    /*  Declare and initialize variables.  */
  int i=0, digit, num_dig=max_ch-1;

    /*  See if x is too large to convert.  */
    if (x >= pow(10,max_ch))
      return 0;
    if (x < 0)
        if (-x >= pow(10,max_ch-1))
          return 0;
        else
        {
            /* Not too big, but a minus is required */
            ch[i++] = '-';
            x = -x;
        }

    /* Now convert it */
    while(i < max_ch)
    {
        digit = x / (pow(10,num_dig-i));
        x = x - (digit*pow(10,num_dig-i));
        ch[i++] = digit + '0';
    }

    /*  Return value to indicate correct conversion.  */
    return 1;
}
/*------------------------------------------------------------------*/

/*------------------------------------------------------------------*/
/*  Function chapter7_27                                            */
/*                                                                  */
/*  This function converts characters in an integer array          */
/*  to an integer numerical value.  If the max_ch number           */
/*  of characters cannot be converted correctly to an integer,     */
/*  the function returns a 0. If successful, the function          */
/*  returns the integer.                                           */

int char_to_int(int ch[], int max_ch)
{
    /*  Declare and initialize variables.  */
    int i, x=0;

    /*  The ch[0] character may be a sign, so do that last.  */
    for (i=1; i<max_ch; i++)
        /* Check if ch[i] is really a digit, and if not, return failure */
        if (isdigit(ch[i]))
            x = x + pow(10,max_ch-i-1)*(ch[i] - '0');
        else
            /* Not a digit, so don't convert, just fail! */
            return 0;
```

```
    /*  Now check the most significant digit, which could be a sign.  */
    switch (ch[0])
    {
        case '-':
            if (x == 0)
                return x;
            else
                return -x;
        case '+':
            return x;
        default:
            /* It wasn't a sign, so check to see if it is a digit */
            if (isdigit(ch[0]))
                return x + pow(10,max_ch-1)*(ch[0] - '0');
            else
                return 0;
    }
}
/*-----------------------------------------------------------------*/


/*-----------------------------------------------------------------*/
/*  Function chapter7_28                                           */
/*                                                                 */
/*  This function receives a pointer to a character string and a   */
/*  character.  The function returns the number of times that the  */
/*  character occurred in the string.                              */

int charcnt(char *ptr, char c)
{
    /*  Declare and initialize variables.  */
    int occur=0, i, length;

    /*  Get the length only once.  */
    length = strlen(ptr);

    /*  Count the occurances of the letter.  */
    for (i=0; i<length; i++)
        if (ptr[i] == c)
            occur++;

    /*  Return occurrences.  */
    return occur;
}
/*-----------------------------------------------------------------*/


/*-----------------------------------------------------------------*/
/*  Function chapter7_29                                           */
/*                                                                 */
/*  This function receives a pointer to a character string and     */
/*  returns the number of repeated characters that occur in        */
/*  the string.  It ignores repeated blanks in the string.  If a   */
/*  character occurs more than once in the string, it is counted   */
/*  as one repeated character.                                     */

int repeat(char *ptr)
{
    /*  Declare and initialize variables.  */
    int i, length, repeaters=0, allchars[NUM_CHAR];
```

```
   /*  Initialize the array of character counts.  */
   for (i=0; i<NUM_CHAR; i++)
      allchars[i] = 0;

   /*  Count the number of times a character occurrs.  */
   length = strlen(ptr);
   for(i=0; i<length; i++)
      allchars[ptr[i]]++;

   /*  Ignore control characters and spaces. Start at the ASCII '!'  */
   for (i = (int)'!'; i<NUM_CHAR; i++)
      if (allchars[i] > 1)
         /*  If a character occurs more than once, count it.  */
         repeaters++;

   /*  Return count of repeated characters.  */
   return repeaters;
}
/*------------------------------------------------------------------*/

/*------------------------------------------------------------------*/
/*  Function chapter7_30                                            */
/*                                                                  */
/*  This function receives a pointer to a character string and      */
/*  returns the number of pairs of the same character.  It          */
/*  ignores repeated blanks in the string.                          */

int repeat2(char *ptr)
{
   /*  Declare and initialize variables.  */
   int i, length, repeaters=0, allchars[NUM_CHAR];

   /*  Initialize the array of character counts.  */
   for (i=0; i<NUM_CHAR; i++)
      allchars[i] = 0;

   /*  Count the number of times a character occurrs.  */
   length = strlen(ptr);
   for(i=0; i<length; i++)
      allchars[ptr[i]]++;

   /*  Ignore control characters and spaces. Start at the ASCII '!'  */
   for (i=(int)'!'; i<NUM_CHAR; i++)
      if (allchars[i] > 1)
         repeaters += allchars[i]-1;

   /*  Return the number of pairs of the same character.  */
   return repeaters;
}
/*------------------------------------------------------------------*/

/*------------------------------------------------------------------*/
/*  Function chapter7_31                                            */
/*                                                                  */
/*  This function receives pointers to two characters string        */
/*  and returns a count of the number of times that the second      */
/*  string occurs in the first.  It does not allow overlap of       */
/*  the occurrences.                                                */

int pattern(char *ptr1, char *ptr2)
```

```
{
    /*  Declare and initialize variables.   */
    int occurances=0;
    char *tmp, arraytmp[MAXLENGTH];

    /*  Copy the first string to a temporary variable.    */
    /*  Otherwise, a side effect of this function would    */
    /*  be destruction of ptr1.                            */
    tmp = arraytmp;
    strcpy(tmp,ptr1);

    /*  Repeat until no more copies of the pattern in the string.  */
    do
    {
        /*  Get a pointer to the beginning of the pattern.  */
        tmp = strstr(tmp,ptr2);

        /*  Check to see if a pattern was found.  */
        if (tmp != NULL)
        {
            occurances++;
            /*  Remove the part of the string already searched.  */
            strcpy(tmp,&tmp[strlen(ptr2)]);
        }
    } while (tmp != NULL);

    /*  Return a count of number of non-overlapping patterns.  */
    return occurances;
}
/*------------------------------------------------------------------*/

/*------------------------------------------------------------------*/
/*  Function chapter7_32                                            */
/*                                                                  */
/*  This function receives pointers to two characters string        */
/*  and returns a count of the number of times that the second      */
/*  string occurs in the first.  It allows overlap of occurrences.  */

int overlap(char *ptr1, char *ptr2)
{
    /*  Declare and initialize variables.   */
    int occurances=0;
    char *tmp, arraytmp[MAXLENGTH];

    /*  Copy the first string to a temporary variable.    */
    /*  Otherwise, a side effect of this function would    */
    /*  be destruction of ptr1.                            */
    tmp = arraytmp;
    strcpy(tmp,ptr1);

    /*  Repeat until no more copies of the pattern in the string.  */
    do
    {
        /*  Get a pointer to the beginning of the pattern.  */
        tmp = strstr(tmp,ptr2);

        /*  Check to see if a pattern was found.  */
        if (tmp != NULL)
        {
            occurances++;
```

```
                /*  Remove the part of the string already searched.  */
                strcpy(tmp, &tmp[1]);
            }
        } while (tmp != NULL);

        /*  Return a count of number of non-overlapping patterns.  */
        return occurances;
}
/*-------------------------------------------------------------------*/
```

Fig. 1.1 Internal Organization of a computer

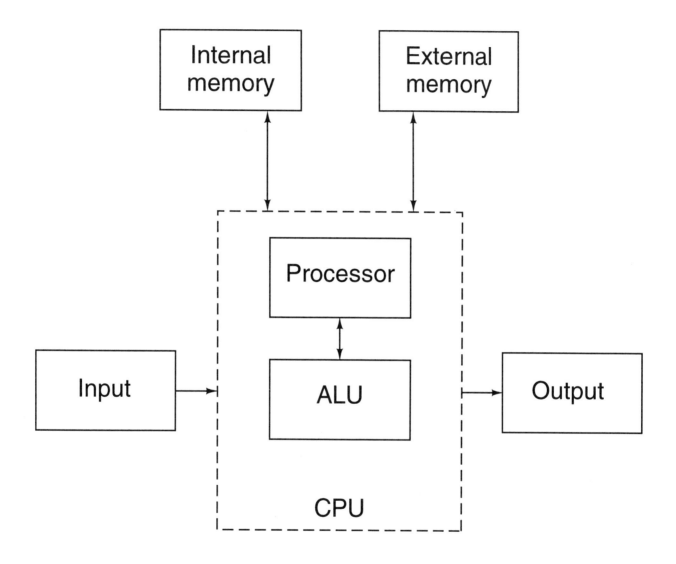

ENGINEERING PROBLEM SOLVING with ANSI C:
FUNDAMENTAL CONCEPTS
by Etter

Fig. 1.2 Software interface to the computer

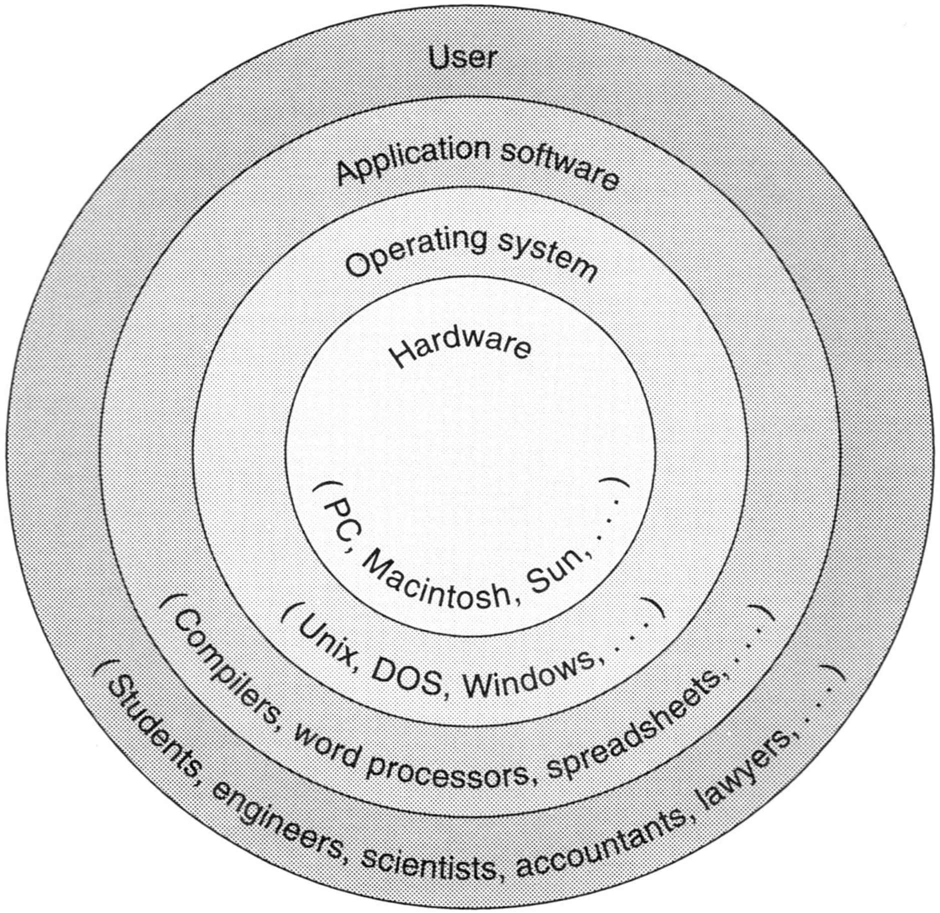

Circles (from outer to inner):
- User (Students, engineers, scientists, accountants, lawyers, ...)
- Application software (Compilers, word processors, spreadsheets, ...)
- Operating system (Unix, DOS, Windows, ...)
- Hardware (PC, Macintosh, Sun, ...)

ENGINEERING PROBLEM SOLVING with ANSI C:
FUNDAMENTAL CONCEPTS
by Etter

Table 1.1 Comparison of software statements

TABLE 1.1 Comparison of Software Statements

Software	Example Statement
C	`area = 3.141593*(diameter/2) * (diameter/2);`
MATLAB	`area = pi*((diameter/2)^2);`
Fortran	`area = 3.141593*(diameter/2.0)**2`
Ada	`area := 3.141593*(diameter/2)**2;`
Pascal	`area := 3.141593*(diameter/2)*(diameter/2)`
Basic	`let a = 3.141593*(d/2)*(d/2)`
COBOL	`compute area = 3.141593*(diameter/2)*(diameter/2).`

ENGINEERING PROBLEM SOLVING with ANSI C:
FUNDAMENTAL CONCEPTS
by Etter

Fig. 1.3   Program compilation/linking/execution

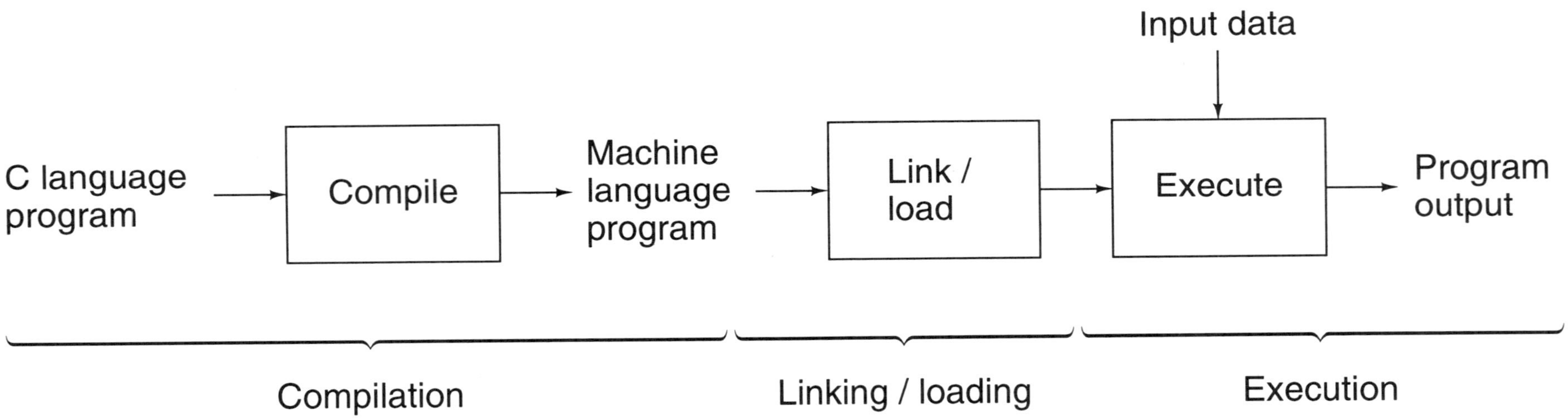

Compilation

Linking / loading

Execution

C language program → Compile → Machine language program → Link / load → Execute → Program output

Input data →

ENGINEERING PROBLEM SOLVING with ANSI C:
FUNDAMENTAL CONCEPTS
by Etter

Table 1.2 Software life-cycle phases

TABLE 1.2 Software Life-Cycle Phases

Life Cycle	Percent of Effort
Definition	3
Specification	15
Coding and modular testing	14
Integrated testing	8
Maintenance	60

ENGINEERING PROBLEM SOLVING with ANSI C:
FUNDAMENTAL CONCEPTS
by Etter

Variable declaration and initialization

```
double x1=1, y1=5, x2=4, y2=7,
       side_1, side_2, distance;
```

x1   `1`

y1   `5`

x2   `4`

y2   `7`

side_1   `?`

side_2   `?`

distance   `?`

ENGINEERING PROBLEM SOLVING with ANSI C:
FUNDAMENTAL CONCEPTS
by Etter

TABLE 2.1 Keywords

auto	double	ints	truct
break	else	long	switch
case	enum	register	typedef
char	extern	return	union
const	float	short	unsigned
continue	for	signed	void
default	goto	sizeof	volatile
do	if	static	while

ENGINEERING PROBLEM SOLVING with ANSI C:
FUNDAMENTAL CONCEPTS
by Etter

Numeric data types

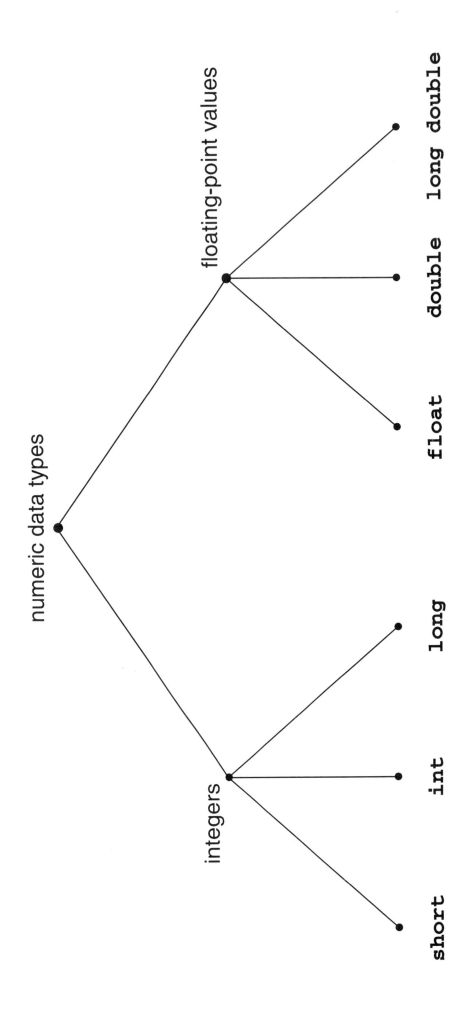

ENGINEERING PROBLEM SOLVING with ANSI C:
FUNDAMENTAL CONCEPTS
by Etter

TABLE 2.2 Example Data-Type Limits*	
Integers	
short	Maximum = 32,767
int	Maximum = 32,767
long	Maximum = 2,147,483,647
Floating Point	
float	6 digits of precision Maximum exponent = 38 Maximum value = $3.402823e + 38$
double	15 digits of precision Maximum exponent = 308 Maximum value = $1.797693e + 308$
long double	19 digits of precision Maximum exponent = 4932 Maximum value = $1.189731e + 4932$

*Borland Turbo C++ 3.0 compiler.

ENGINEERING PROBLEM SOLVING with ANSI C:
FUNDAMENTAL CONCEPTS
by Etter

Precedence of arithmetic operators

TABLE 2.3 Precedence of Arithmetic Operators

Precedence	Operator	Associativity
1	Parentheses: ( )	Innermost first
2	Unary operators: + − (type)	Right to left
3	Binary operators: * / %	Left to right
4	Binary operators: + −	Left to right

ENGINEERING PROBLEM SOLVING with ANSI C:
FUNDAMENTAL CONCEPTS
by Etter

TABLE 2.4 Precedence of Arithmetic and Assignment Operators

Precedence	Operator	Associativity
1	Parentheses: ( )	Innermost first
2	Unary operators: + - ++ -- (type)	Right to left
3	Binary operators: * / %	Left to right
4	Binary operators: + -	Left to right
5	Assignment operators: = += -= *= /= %=	Right to left

ENGINEERING PROBLEM SOLVING with ANSI C:
FUNDAMENTAL CONCEPTS
by Etter

Numeric conversion specifiers for output statements

TABLE 2.5 Numeric Conversion Specifiers for Output Statements

Variable Type	Output Type	Specifier
Integer Values		
short, int	int	%i, %d
int	short	%hi, %hd
long	long	%li, %ld
int	unsigned int	%u
int	unsigned short	%hu
long	unsigned long	%lu
Floating-Point Values		
float, double	double	%f, %e, %E, %g, %G
long double	long double	%Lf, %Le, %LE, %Lg, %LG

ENGINEERING PROBLEM SOLVING with ANSI C:
FUNDAMENTAL CONCEPTS
by Etter

Examples with conversion specifiers

Specifier	Value Printed
%i	-145
%4d	-145
%3i	-145
%6i	bb-145
%-6i	-145bb

Specifier	Value Printed
%f	157.892600
%6.2f	157.89
%+8.2f	b+157.89
%7.5f	157.89260
%e	1.578926e+02
%.3E	1.579E+02
%g	157.893

ENGINEERING PROBLEM SOLVING with ANSI C:
FUNDAMENTAL CONCEPTS
by Etter

Escape sequences

sequence	character represented
\a	alert(bell) character
\b	backspace
\f	formfeed
\n	newline
\r	carriage return
\t	horizontal tab
\v	vertical tab
\\	backslash
\?	question mark
\'	single quote
\"	double quote

ENGINEERING PROBLEM SOLVING with ANSI C:
FUNDAMENTAL CONCEPTS
by Etter

Numeric conversion specifiers for input statements

TABLE 2.6 Numeric Conversion Specifiers
for Input Statements

Variable Type	Specifier
Integer Values	
int	%i, %d
short	%hi, %hd
long int	%li, %ld
unsigned int	%u
unsigned short	%hu
unsigned long	%lu
Floating-Point Values	
float	%f, %e, %E, %g, %G
double	%lf, %le, %lE, %lg, %lG
long double	%Lf, %Le, %LE, %Lg, %LG

ENGINEERING PROBLEM SOLVING with ANSI C:
FUNDAMENTAL CONCEPTS
by Etter

Fig. 2.1 Linear and cubic spline interpolation

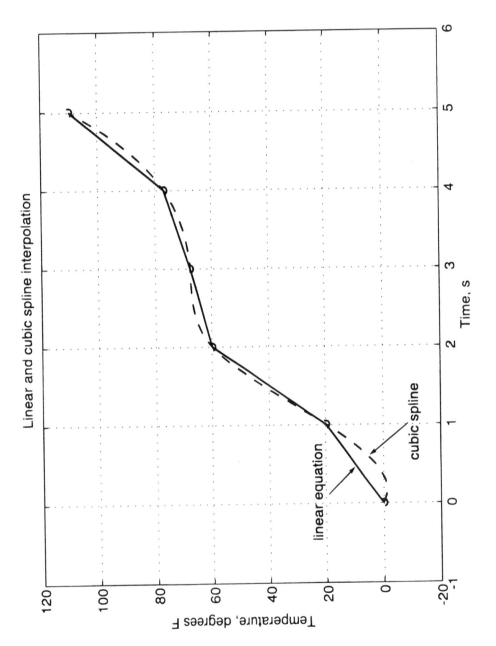

Linear and cubic spline interpolation

ENGINEERING PROBLEM SOLVING with ANSI C:
FUNDAMENTAL CONCEPTS
by Etter

Fig. 2.2 Linear interpolation using similar triangles

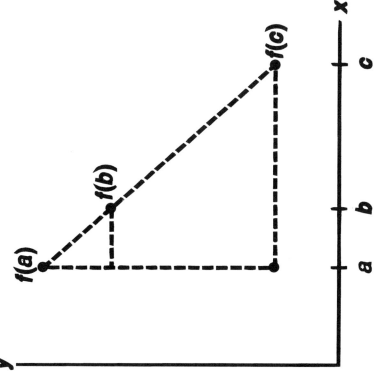

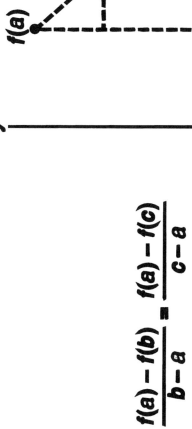

$$\frac{f(a) - f(b)}{b - a} = \frac{f(a) - f(c)}{c - a}$$

ENGINEERING PROBLEM SOLVING with ANSI C:
FUNDAMENTAL CONCEPTS
by Etter

Fig. 3.1   Pseudocode notation and flowchart symbols

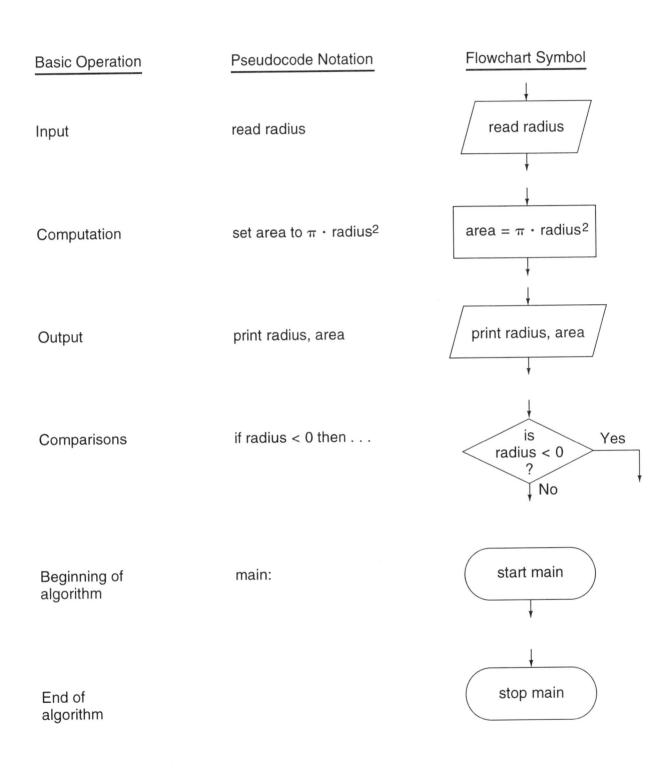

ENGINEERING PROBLEM SOLVING with ANSI C:
FUNDAMENTAL CONCEPTS
by Etter

Fig. 3.2 Example of a flowchart for a sequence structure

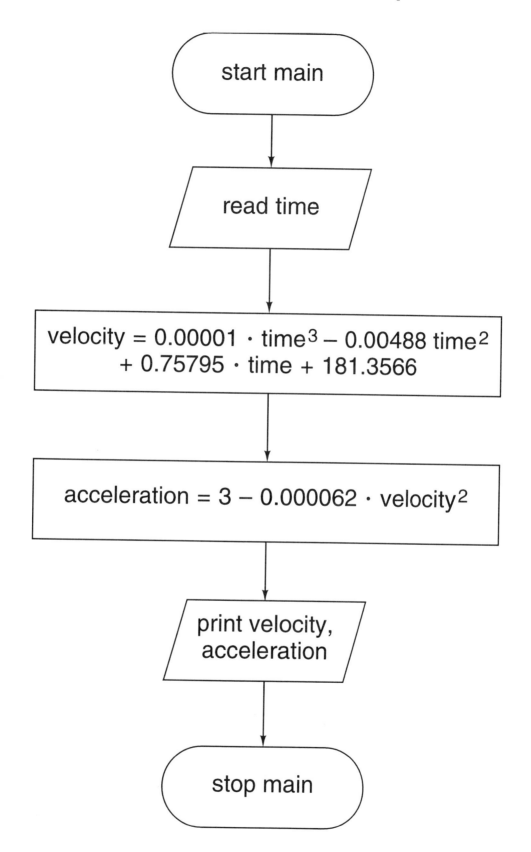

Fig. 3.3   Example of a flowchart for a selection structure

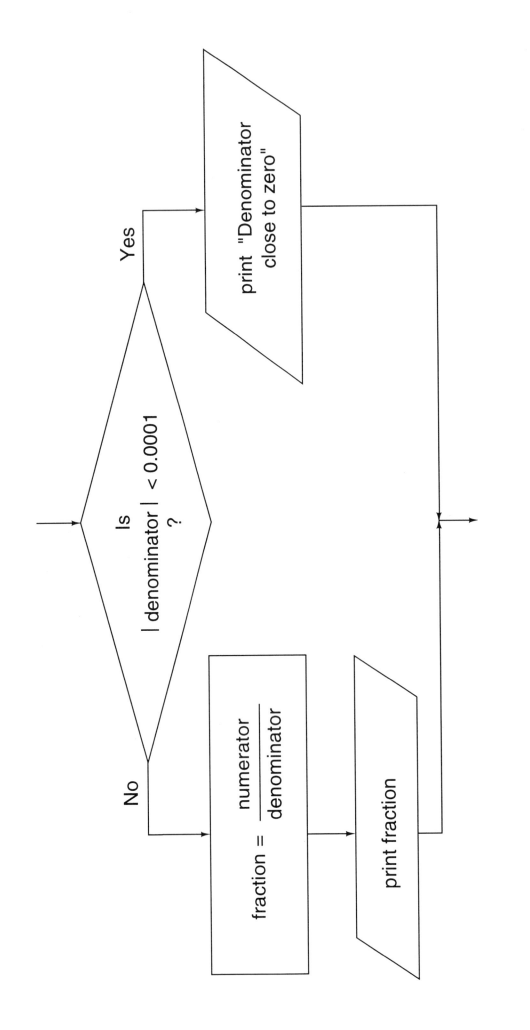

ENGINEERING PROBLEM SOLVING with ANSI C:
FUNDAMENTAL CONCEPTS
by Etter

Fig. 3.4 Example of a flowchart for a repetition structure

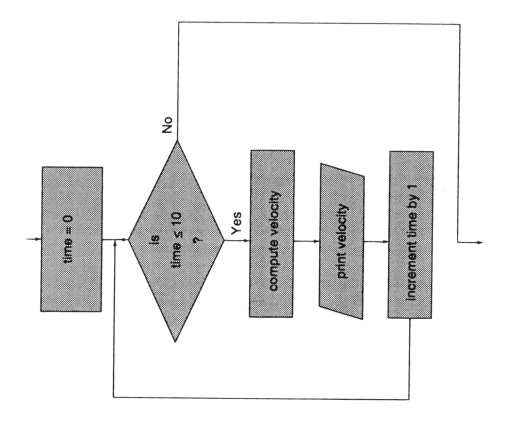

ENGINEERING PROBLEM SOLVING with ANSI C:
FUNDAMENTAL CONCEPTS
by Etter

Relational Operator

Relational Operator	Interpretation
<	is less than
<=	is less than or equal to
>	is greater than
>=	is greater than or equal to
==	is equal to
!=	is not equal to

ENGINEERING PROBLEM SOLVING with ANSI C:
FUNDAMENTAL CONCEPTS
by Etter

Logical operators

TABLE 3.1 Logical Operators

A	B	A && B	A || B	!A	!B
False	False	False	False	True	True
False	True	False	True	True	False
True	False	False	True	False	True
True	True	True	True	False	False

ENGINEERING PROBLEM SOLVING with ANSI C:
FUNDAMENTAL CONCEPTS
by Etter

Operator precedence for arithmetic, relational and logical operators

TABLE 3.2 Operator Precedence for Arithmetic, Relational, and Logical Operators

Precedence	Operation	Associativity
1	( )	innermost first
2	++ -- + - ! (type)	right to left (unary)
3	* / %	left to right
4	+ -	left to right
5	< <= > >=	left to right
6	== !=	left to right
7	&&	left to right
8	||	left to right
9	= += -= *= /= %=	right to left

ENGINEERING PROBLEM SOLVING with ANSI C:
FUNDAMENTAL CONCEPTS
by Etter

Fig. 3.9 Distances from a linear estimate to a set of points

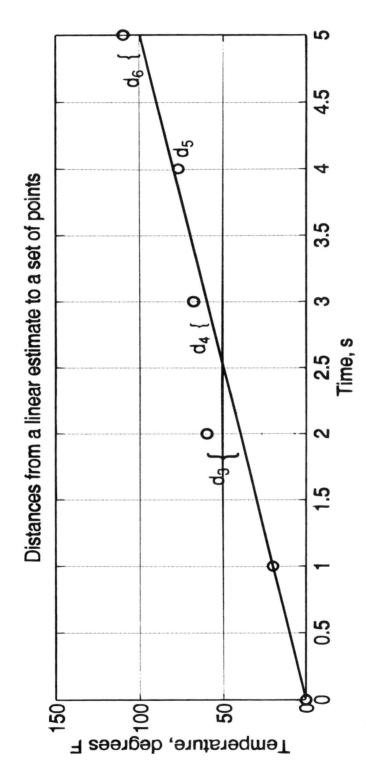

Distances from a linear estimate to a set of points

ENGINEERING PROBLEM SOLVING with ANSI C:
FUNDAMENTAL CONCEPTS
by Etter

Equations for the "best" linear fit in terms of least squares

$$m = \frac{\sum_{k=1}^{n} x_k \cdot \sum_{k=1}^{n} y_k - n \cdot \sum_{k=1}^{n} x_k y_k}{\left(\sum_{k=1}^{n} x_k\right)^2 - n \cdot \sum_{k=1}^{n} x_k^2}$$

$$b = \frac{\sum_{k=1}^{n} x_k \cdot \sum_{k=1}^{n} x_k y_k - \sum_{k=1}^{n} x_k^2 \cdot \sum_{k=1}^{n} y_k}{\left(\sum_{k=1}^{n} x_k\right)^2 - n \cdot \sum_{k=1}^{n} x_k^2}$$

ENGINEERING PROBLEM SOLVING with ANSI C:
FUNDAMENTAL CONCEPTS
by Etter

Equations for the "best" linear fit in terms of least squares

The best fit to a set of points computed using linear regression

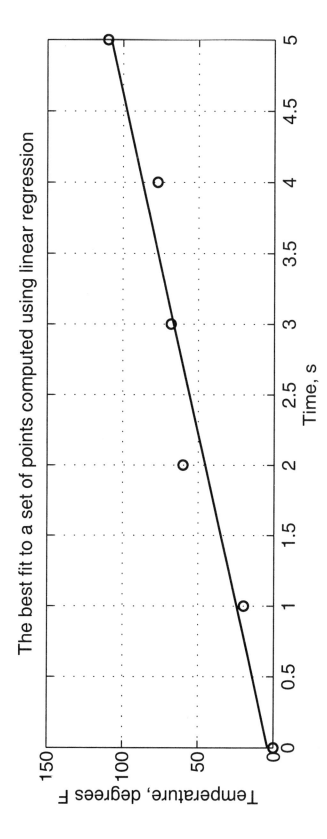

© 1995 by Prentice-Hall, Inc.
A Simon & Schuster Company
Englewood Cliffs, NJ 07632

ENGINEERING PROBLEM SOLVING with ANSI C:
FUNDAMENTAL CONCEPTS
by Etter

Fig. 3.11 Atmospheric layers around the earth

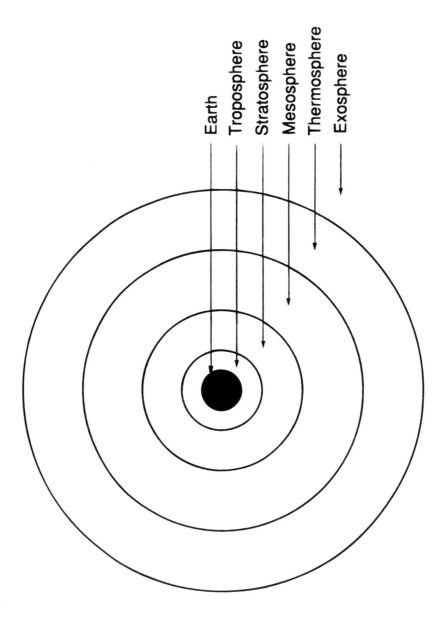

Earth

Troposphere

Stratosphere

Mesosphere

Thermosphere

Exosphere

ENGINEERING PROBLEM SOLVING with ANSI C:
FUNDAMENTAL CONCEPTS
by Etter

Fig. 4.1 Example structure charts

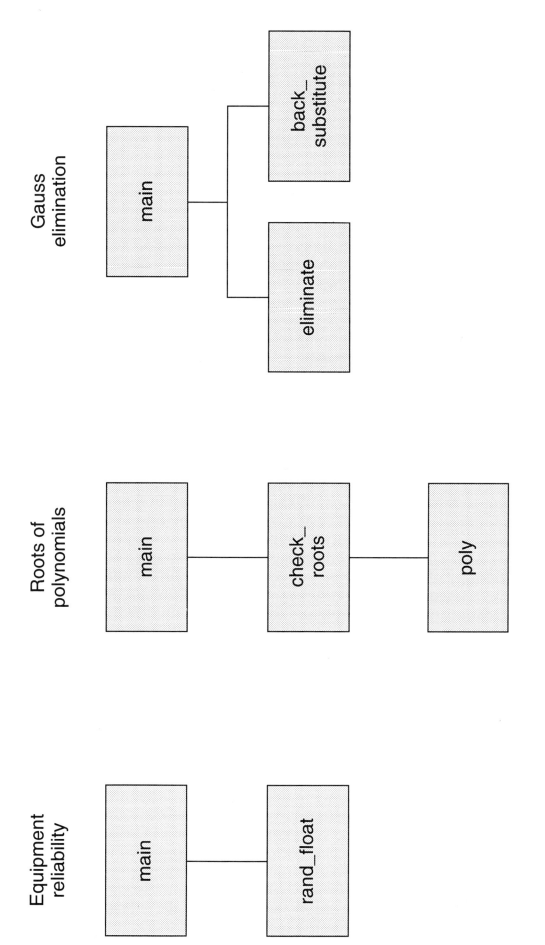

ENGINEERING PROBLEM SOLVING with ANSI C:
FUNDAMENTAL CONCEPTS
by Etter

Fig. 4.1 Example structure chart

Speech signal
analysis

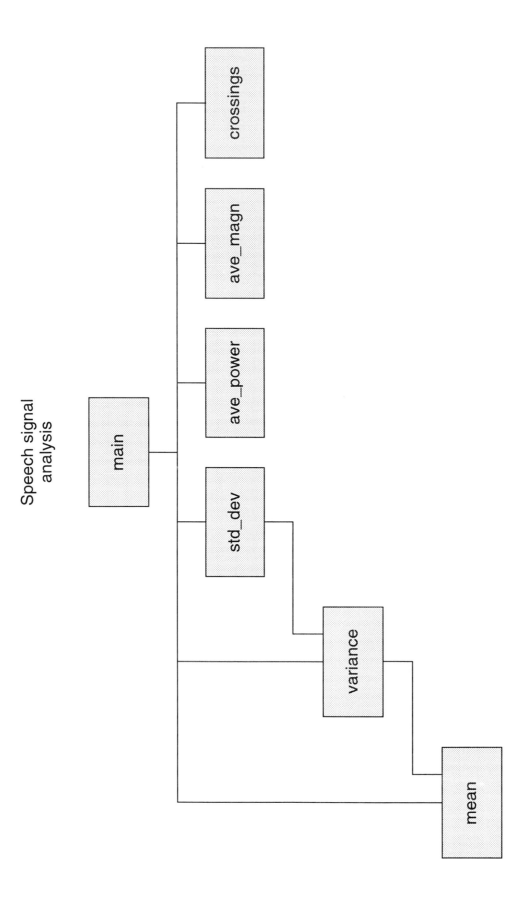

Fig. 4.4 Series and parallel configurations

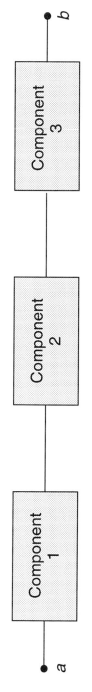

(a) Series design.

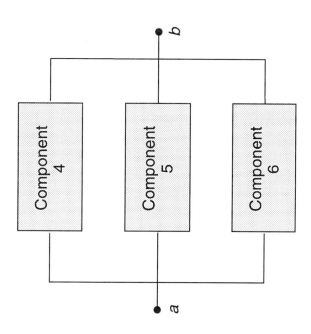

(b) Parallel design.

ENGINEERING PROBLEM SOLVING with ANSI C:
FUNDAMENTAL CONCEPTS
by Etter

Fig. 4.6 Cubic polynomials

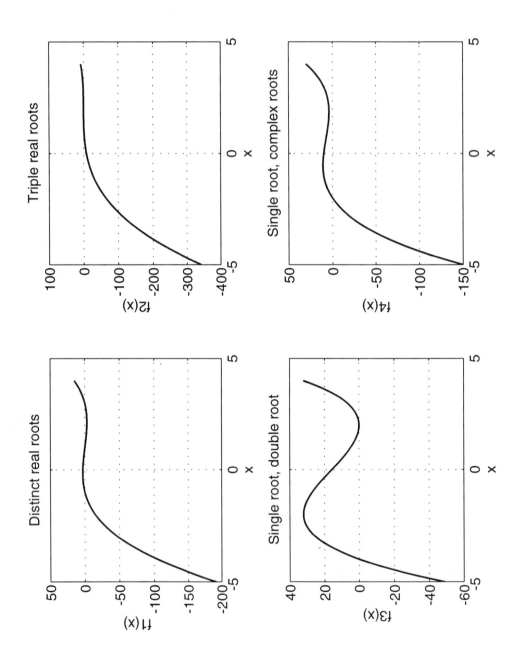

ENGINEERING PROBLEM SOLVING with ANSI C:
FUNDAMENTAL CONCEPTS
by Etter

Fig. 4.7 Incremental search for roots

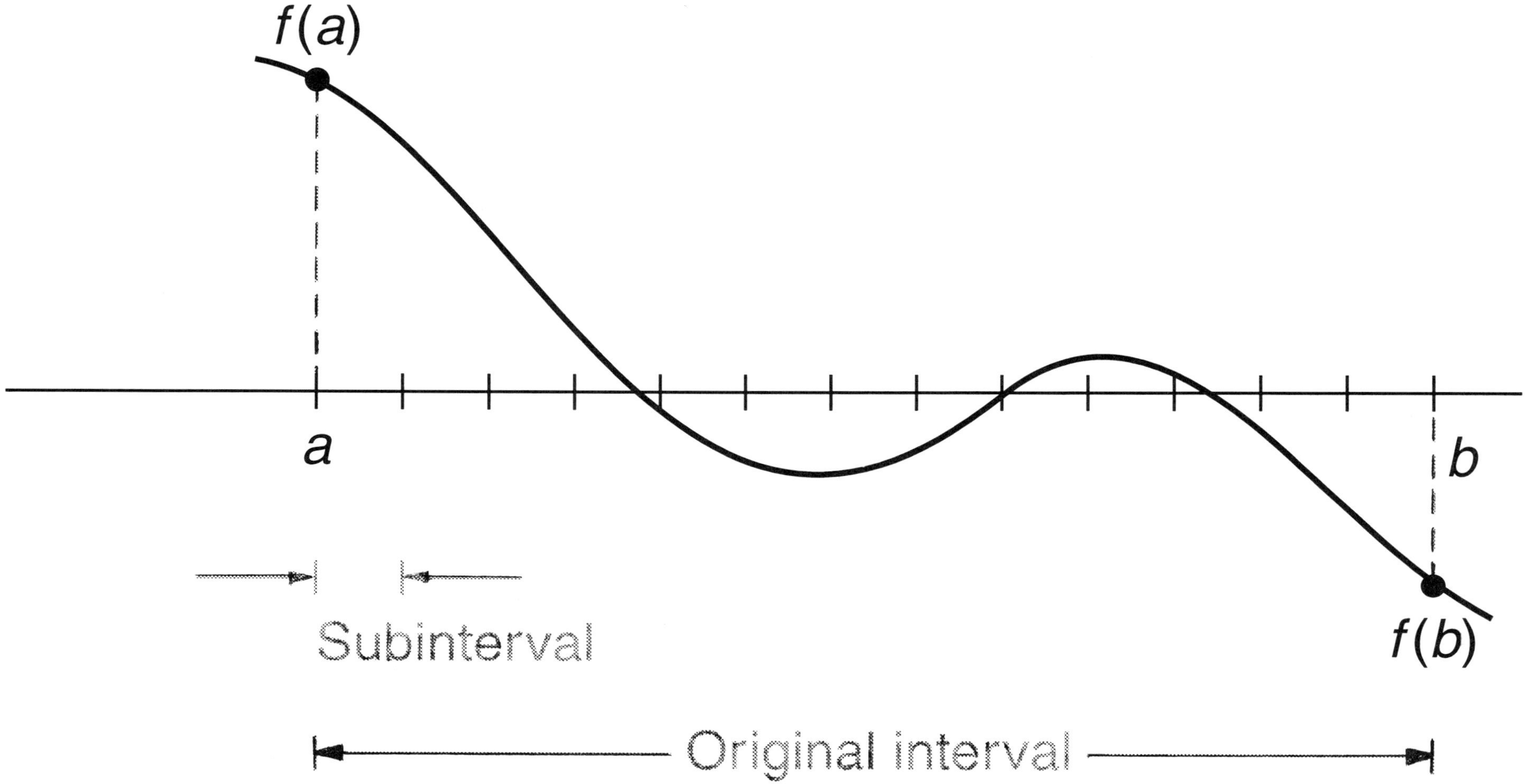

$f(a)$

a

b

$f(b)$

Subinterval

Original interval

ENGINEERING PROBLEM SOLVING with ANSI C:
FUNDAMENTAL CONCEPTS
by Etter

Fig. 4.8 (a) Subinterval analysis

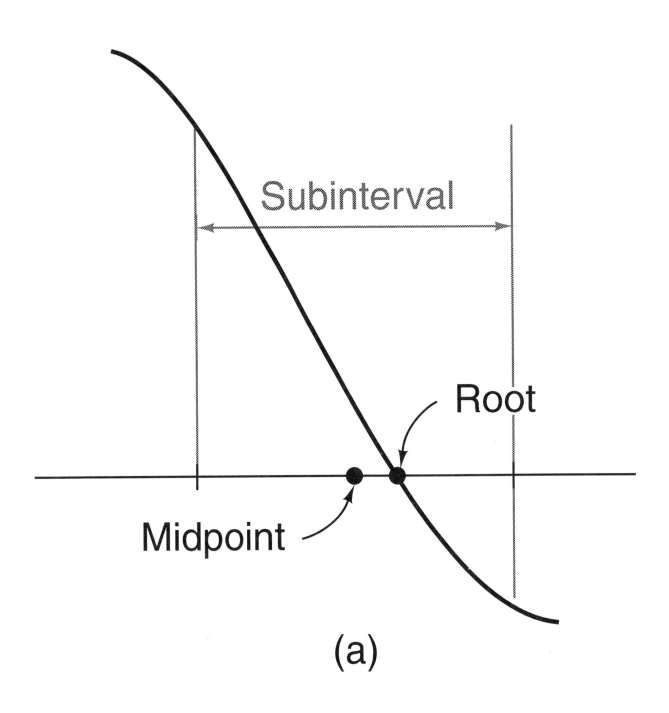

(a)

ENGINEERING PROBLEM SOLVING with ANSI C:
FUNDAMENTAL CONCEPTS
by Etter

© 1995 by Prentice-Hall, Inc.
A Simon & Schuster Company
Englewood Cliffs, NJ 07632

Fig. 4.8(b) Subinterval analysis

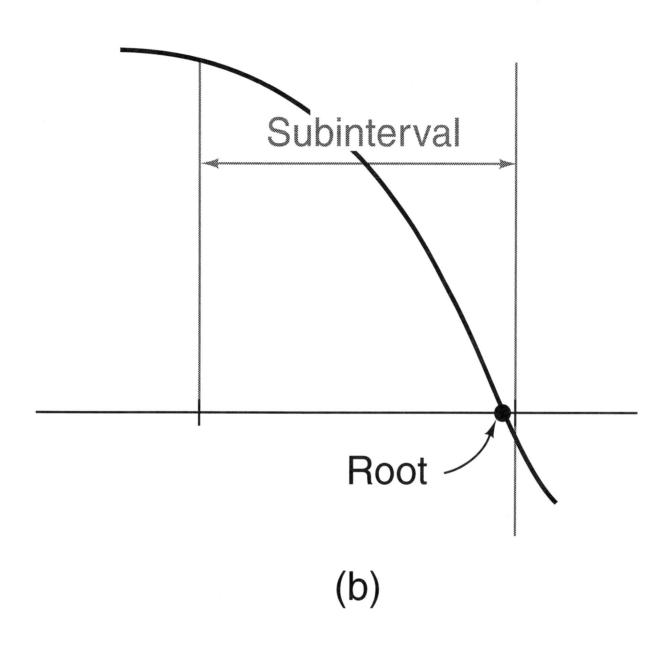

Subinterval

Root

(b)

ENGINEERING PROBLEM SOLVING with ANSI C:
FUNDAMENTAL CONCEPTS
by Etter

One-dimensional arrays

5	0	-1	2	15	2
s[0]	s[1]	s[2]	s[3]	s[4]	s[5]

0.0	t[0]
0.1	t[1]
0.2	t[2]
0.3	t[3]

© 1995 by Prentice-Hall, Inc.
A Simon & Schuster Company
Englewood Cliffs, NJ 07632

ENGINEERING PROBLEM SOLVING with ANSI C:
FUNDAMENTAL CONCEPTS
by Etter

Table 5.1 Operator precedence

TABLE 5.1 Operator Precedence

Precedence	Operation	Associativity
1	( ) [ ]	innermost first
2	++ -- + - ! (type)	right to left (unary)
3	* / %	left to right
4	+ -	left to right
5	< <= > >=	left to right
6	== !=	left to right
7	&&	left to right
8	||	left to right
9	? :	right to left
10	= += -= *= /= %=	right to left
11	,	left to right

ENGINEERING PROBLEM SOLVING with ANSI C :
FUNDAMENTAL CONCEPTS
by Etter

Fig. 5.1 Random sequences

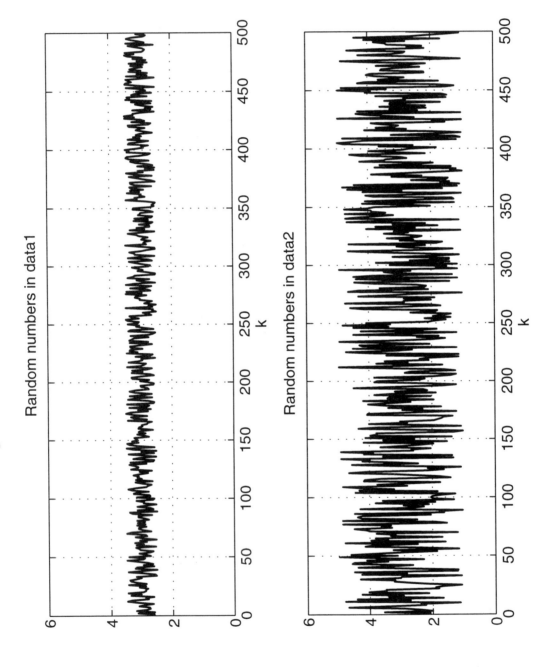

Random numbers in data1

Random numbers in data2

ENGINEERING PROBLEM SOLVING with ANSI C:
FUNDAMENTAL CONCEPTS
by Etter

Fig. 5.2 Utterance of the word "zero"

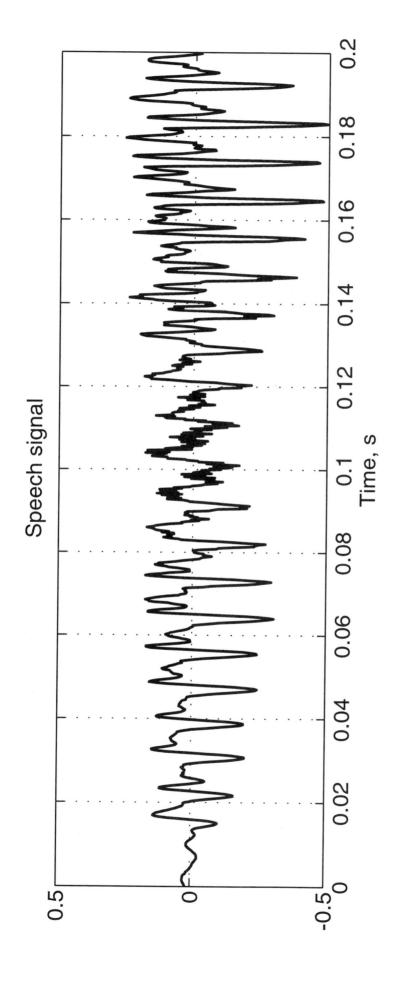

Speech signal

Time, s

ENGINEERING PROBLEM SOLVING with ANSI C:
FUNDAMENTAL CONCEPTS
by Etter

Two-dimensional array

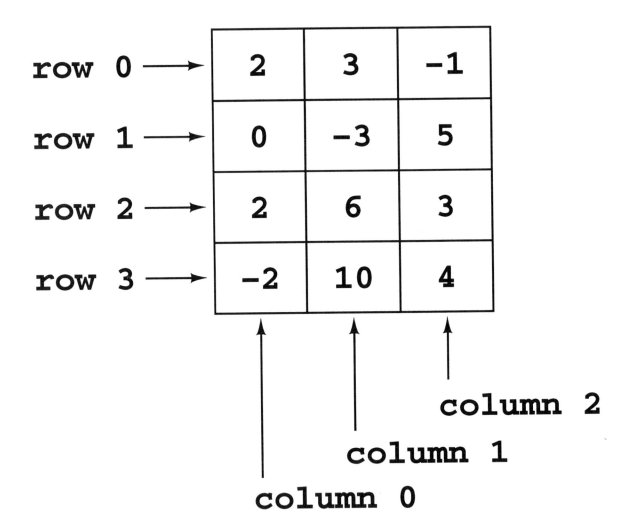

Fig. 5.3(a) Two lines

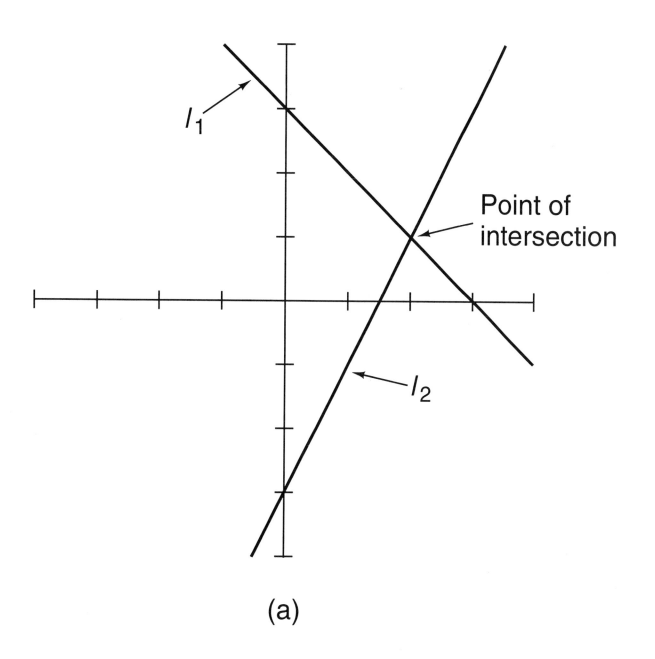

(a)

ENGINEERING PROBLEM SOLVING with ANSI C:
FUNDAMENTAL CONCEPTS
by Etter

Fig. 5.3(b) Two lines

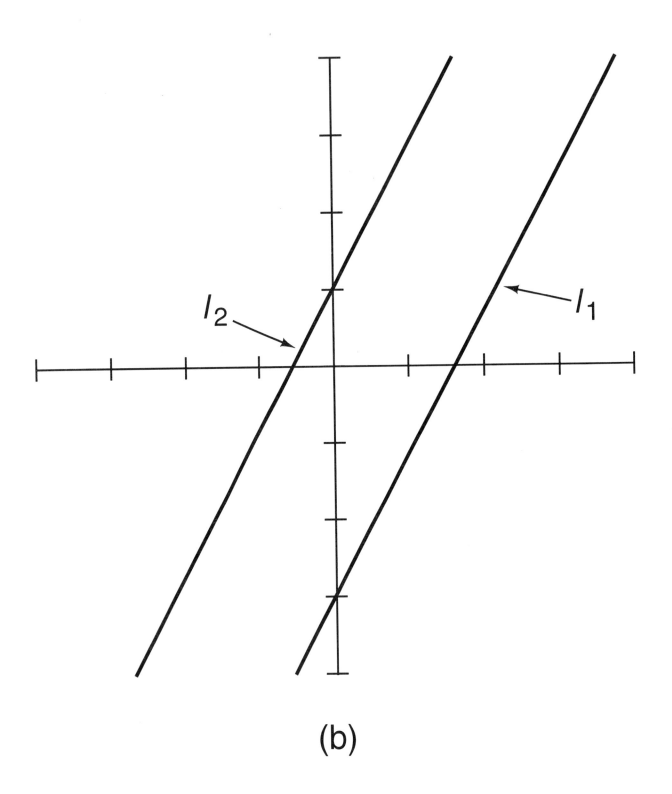

(b)

ENGINEERING PROBLEM SOLVING with ANSI C:
FUNDAMENTAL CONCEPTS
by Etter

Fig. 5.3(c) Two lines

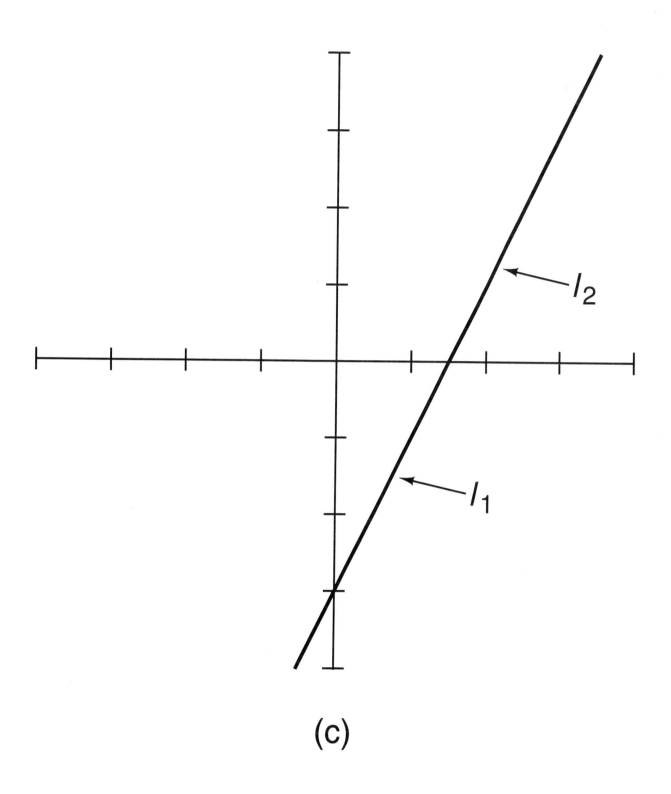

(c)

ENGINEERING PROBLEM SOLVING with ANSI C:
FUNDAMENTAL CONCEPTS
by Etter

Fig. 5.4(a) Two planes

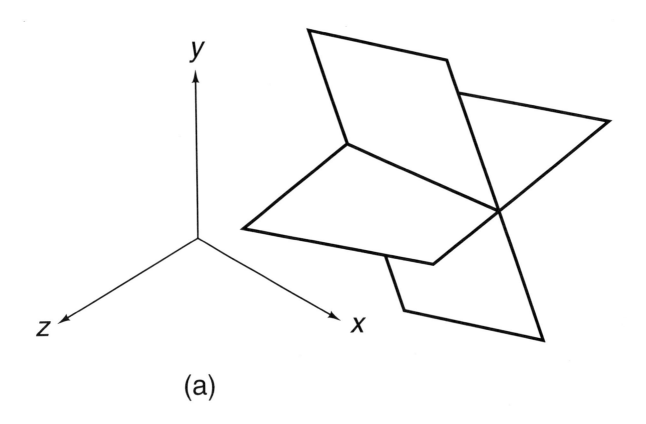

(a)

Fig. 5.4(b) Two planes

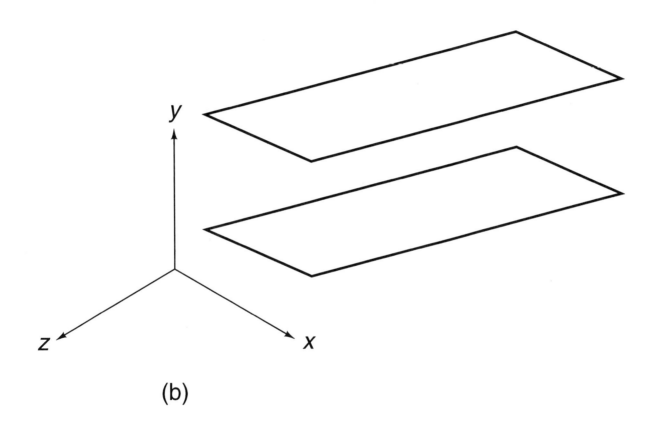

(b)

ENGINEERING PROBLEM SOLVING with ANSI C:
FUNDAMENTAL CONCEPTS
by Etter

Fig. 5.4(c) Two planes

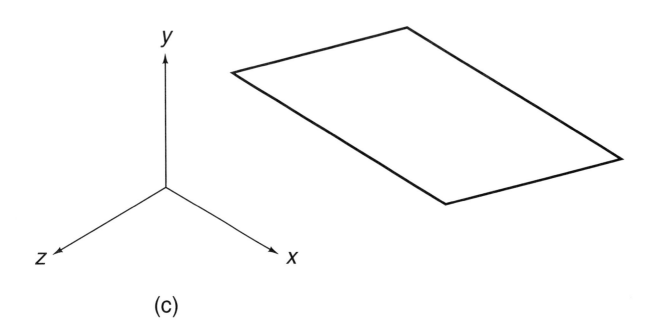

(c)

ENGINEERING PROBLEM SOLVING with ANSI C:
FUNDAMENTAL CONCEPTS
by Etter

© 1995 by Prentice-Hall, Inc.
A Simon & Schuster Company
Englewood Cliffs, NJ 07632

Fig. 5.5(a) Three distinct planes

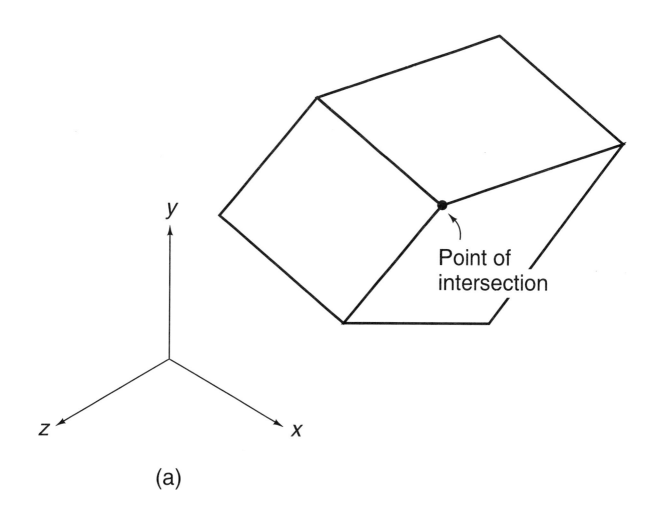

Point of
intersection

(a)

ENGINEERING PROBLEM SOLVING with ANSI C:
FUNDAMENTAL CONCEPTS
by Etter

Fig. 5.5(b) Three distinct planes

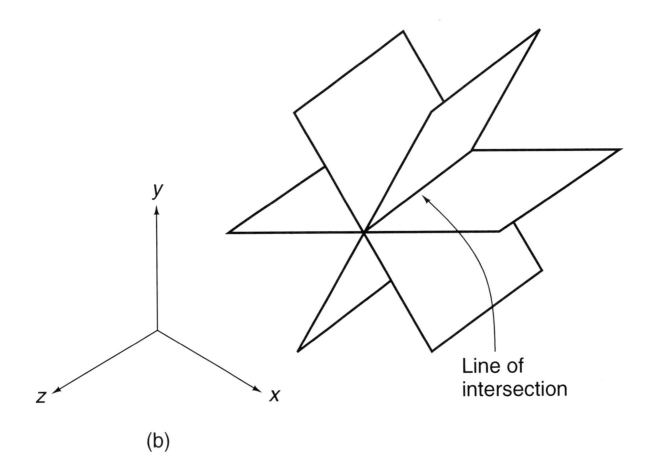

Line of
intersection

(b)

ENGINEERING PROBLEM SOLVING with ANSI C:
FUNDAMENTAL CONCEPTS
by Etter

Fig. 5.5(c) Three distinct planes

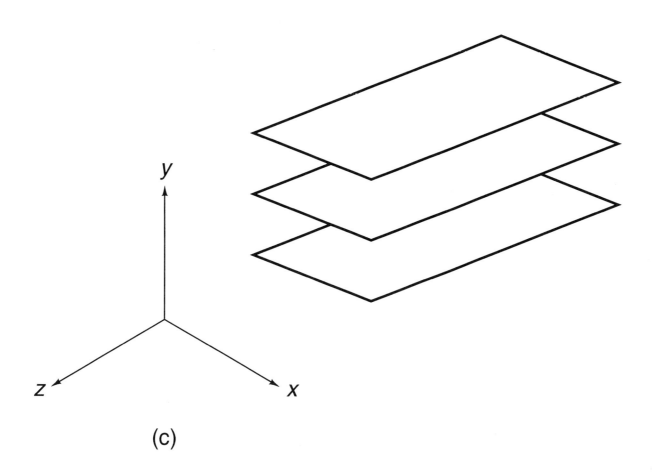

(c)

Fig. 5.5(d) Three distinct planes

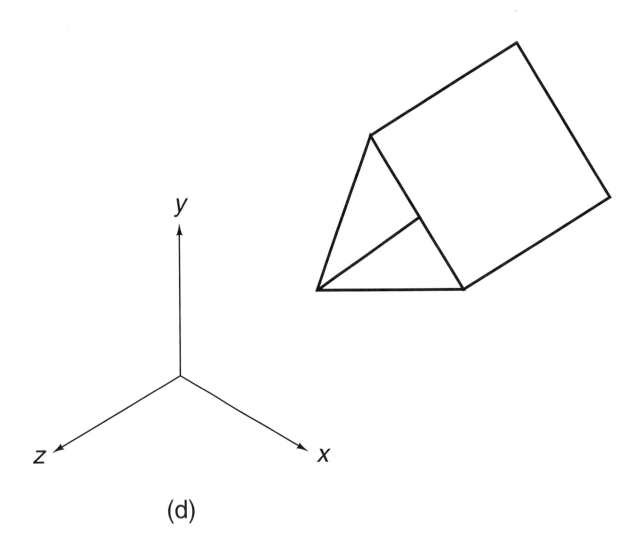

(d)

Fig. 5.5(e) Three distinct planes

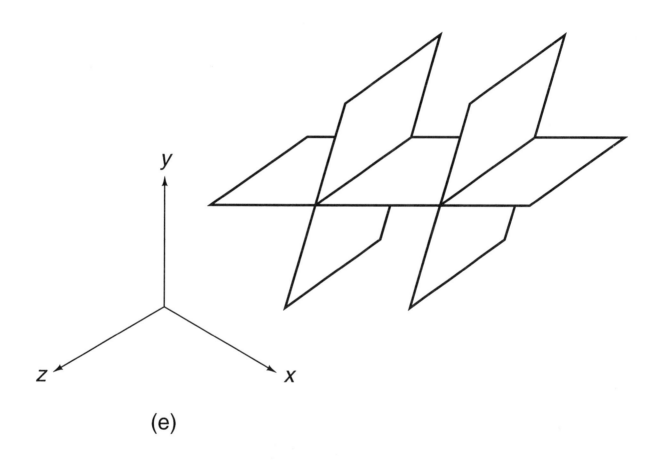

(e)

ENGINEERING PROBLEM SOLVING with ANSI C:
FUNDAMENTAL CONCEPTS
by Etter

Fig. 5.7 Three-dimensional array

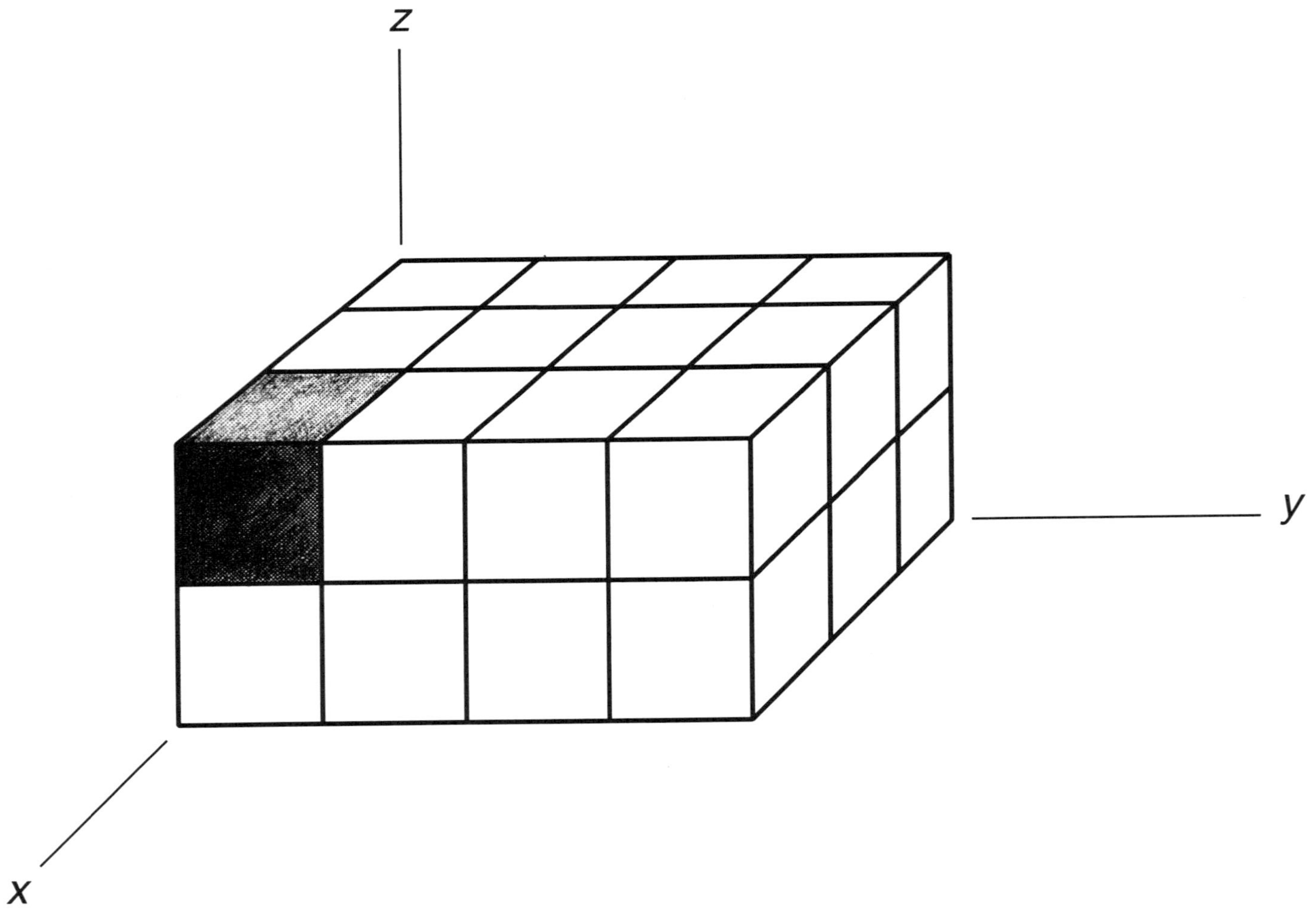

ENGINEERING PROBLEM SOLVING with ANSI C:
FUNDAMENTAL CONCEPTS
by Etter

Analogy of post office boxes to memory allocation

post office box number	individual name	contents
78	John Ruiz	catalog

memory address	identifier	contents
66572	x	105

ENGINEERING PROBLEM SOLVING with ANSI C:
FUNDAMENTAL CONCEPTS
by Etter

Memory snapshot

```
int a, b, *ptr=&a;
```

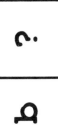

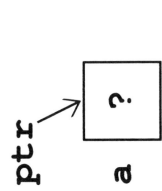

ptr

a ?

b ?

ENGINEERING PROBLEM SOLVING with ANSI C:
FUNDAMENTAL CONCEPTS
by Etter

Memory snapshot

```
/* Declare and initialize variables. */
int a=5, b=9, *ptr=&a;
```

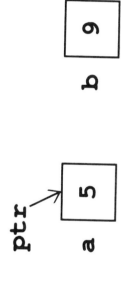

ptr

a 5   b 9

ENGINEERING PROBLEM SOLVING with ANSI C:
FUNDAMENTAL CONCEPTS
by Etter

Memory snapshot

```
/* Declare and initialize variables.  */
int x=-5, y = 8, *ptr_1, *ptr_2;
...
/* Assign both pointers to x.  */
ptr_1 = &x;
ptr_2 = ptr_1;
```

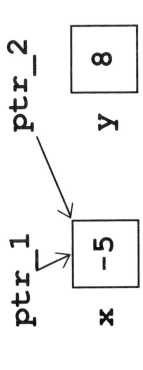

Operator precedence

TABLE 6.1 Operator Precedence

Precedence	Operation	Associativity
1	() []	innermost first
2	++ -- + - ! (type) & *	right to left (unary)
3	* / %	left to right
4	+ -	left to right
5	< <= > >=	left to right
6	== !=	left to right
7	&&	left to right
8	||	left to right
9	? :	right to left
10	= += -= *= /= %=	right to left
11	,	left to right

ENGINEERING PROBLEM SOLVING with ANSI C:
FUNDAMENTAL CONCEPTS
by Etter

One-dimensional array offsets

offset

x[0]	1.5	0
x[1]	2.2	1
x[2]	4.3	2
x[3]	7.5	3
x[4]	9.1	4
x[5]	10.5	5

ENGINEERING PROBLEM SOLVING with ANSI C:
FUNDAMENTAL CONCEPTS
by Etter

Two-dimensional array offsets

2	4	6
1	5	3

offset

		offset
s[0][0]	2	0
s[0][1]	4	1
s[0][2]	6	2
s[1][0]	1	3
s[1][1]	5	4
s[1][2]	3	5

ENGINEERING PROBLEM SOLVING with ANSI C:
FUNDAMENTAL CONCEPTS
by Etter

Examples of ASCII codes

TABLE 7.1 Examples of ASCII Codes

Character	ASCII Code	Integer Equivalent
newline, \n	0001010	10
%	0100101	37
3	0110011	51
A	1000001	65
a	1100001	97
b	1100010	98
c	1100011	99

ENGINEERING PROBLEM SOLVING with ANSI C:
FUNDAMENTAL CONCEPTS
by Etter